T0296124

Cambridge Tracts in Mathematics
and Mathematical Physics

GENERAL EDITOR
G. H. HARDY, M.A., F.R.S.

No. 39

A LOCUS
WITH 25920 LINEAR SELF-
TRANSFORMATIONS

A LOCUS
WITH 25920 LINEAR SELF-TRANSFORMATIONS

BY

H. F. BAKER, Sc.D., LL.D., F.R.S.

Fellow of St. John's College, Cambridge; lately
Lowndean Professor in the University

CAMBRIDGE
AT THE UNIVERSITY PRESS
1946

CAMBRIDGE
UNIVERSITY PRESS

University Printing House, Cambridge CB2 8BS, United Kingdom

Cambridge University Press is part of the University of Cambridge.

It furthers the University's mission by disseminating knowledge in the pursuit of education, learning and research at the highest international levels of excellence.

www.cambridge.org
Information on this title: www.cambridge.org/9781107493711

© Cambridge University Press 1946

First published 1946
Re-issued 2015

A catalogue record for this publication is available from the British Library

ISBN 978-1-107-49371-1 Paperback

CONTENTS

PREFACE

THIS VOLUME is concerned with a locus—itself very interesting to explore geometrically—which exhibits in a simple way the structure of the group of the lines of a cubic surface in ordinary space, regarded as the group of the tritangent planes of the surface. Incidentally certain quite elementary results for the substitutions of 4, 5 and 6 objects are necessary; and, for the sake of completeness, these are explained in detail. Historically the linear expression for the group of transformations considered, and of the different linear expression briefly sketched in Note 2, of the Appendix, both arose from the theory of the linear transformations of the periods of theta functions of two variables; but, beyond references to the literature, this is not dealt with. It is hoped that the Introduction to the text, and the list of headings of the sections, will make sufficiently clear what is included. A brief index of notations is also appended. The argument of the text requires frequent reference to the Scheme of synthemes given as frontispiece of the volume.

To some it may seem that such a theme—at this time—is futile. It is possible, however, to take the view that the primary purpose of the pursuit of science is not the advancement of technology, but the widening of the horizon of the human mind. In this mathematics has always borne an honourable, often a decisive part; indeed, many cases could be cited to support the more extreme view that the development of mathematical ideas, and the emergence of new physical conceptions, are intimately related.

The general theory of linear groups has developed widely on the algebraic side since the group considered here—one of the earliest—was established; and even of this the present is a very

incomplete account. The writer has had the advantage of the
co-operation, in the reading of the proof-sheets, of Dr J. A. Todd,
who has himself written on the matter, and is very greatly in-
debted to him. He would wish also to acknowledge his obligation
to the Staff of the University Press, especially at this time of
difficulty in the printing of books.

H. F. B.

26 *November* 1945

INTRODUCTION

In his monumental volume on the theory of substitutions (*Traité des substitutions*, Paris, 1870), Jordan considers the group of the lines of a cubic surface in ordinary space, which he regards primarily as the group of the substitutions of the tritangent planes of the surface. Later in the same volume, with acknowledgements to Kronecker, he considers the group of the trisection of the periods of a theta function of two variables, proving that the study of this group is essentially the same problem as that of the group of the lines of a cubic surface. In a series of papers* on hyperelliptic functions of two variables, in the *Math. Ann.*, Burkhardt obtains five theta functions which are linearly transformed among themselves by the group of the trisection, thus incidentally obtaining for the first time the expression of the group of the lines of a cubic surface by linear equations (which arises also in a different form in his fourth memoir); and he investigates the homogeneous polynomials in these five functions which are invariant under the resulting linear group. The simplest of these invariants is of the fourth order in the five functions. When equated to zero, this represents a primal in space of four dimensions, which, considering the thoroughness of Burkhardt's work in the four memoirs quoted, may be described as *Burkhardt's Primal*. The geometrical properties of this primal are very interesting; and they form a vivid and simple concrete representation of the group of the lines of a cubic surface, and its more important subgroups; and incidentally illustrate the elements of the theory

* Burkhardt, *Math. Ann.* xxxv (1889), pp. 198–296; xxxvi (1889), pp. 371–434; xxxviii (1890), pp. 161–224, 309–12; xli (1892), pp. 313–43. It is the third of these memoirs xxxviii (1890), which mainly concerns us here, but reference is made to the fourth memoir in Note 2 of the Appendix at the end of this volume.

of the substitutions of five and six objects. After Burkhardt there are two interesting papers by A. B. Coble (*American J. Math.*, XXVIII (1906), pp. 333–66, and *Trans. Amer. Math. Soc.* XVIII (1917), pp. 331–72), in which the geometrical properties of the primal are considered. Coble gives explicitly a symmetrical form of the equation of the primal (to which the equations of transformation are given by Burkhardt, *Math. Ann.* XXVIII, p. 205). On account of its symmetry this form is adopted here as fundamental. Still later Dr J. A. Todd (*Quart. J. Math.* VII (1936), pp. 168–74) has added to his other papers on quartic primals in four dimensions a masterly proof that Burkhardt's primal is rational (that is, that its four independent coordinates are expressible rationally in terms of three rational functions of themselves), without, however, obtaining the explicit reverse equations. It is remarked here that this rationality is obvious when it is seen that there exist on the primal (many) sets of three planes of which every two have only a point in common; and the reverse equations are obtained in one of the possible 72.40 ways.

The present account is primarily a study of the geometrical properties of the primal; and, to be intelligible, must needs contain many results that are not novel. But there are two features which, so far as I know, are new. The first is a notation for the forty-five nodes of the primal (and, therefore, effectively, for the tritangent planes of a cubic surface) which enables the relations of these nodes to be simply described and verified. The second is the reference to the projections into itself of which the primal is capable, of which I have seen no mention. All Burkhardt's fundamental transformations are expressed here in terms of these projections. Burkhardt's proof that these fundamental transformations generate the group depends upon their derivation from linear transformation of the periods of the hyperelliptic functions, and so belongs to the theory of linear transformation of the periods. It would seem that what is advanced below in regard to the geometrical projections is sufficient to enable us to dispense with reference to these periods; but a formal proof of this requires further elaboration. The elementary results which arise for the substitutions of five and six objects will not be new to those who have studied the theory of groups of substitutions, but may be welcome, from their concrete character, to those less familiar with the theory.

One remark should perhaps be added here to make the general statements of this introduction more precise: The group of the lines of a cubic surface is of order $2^4 . 3^4 . 40$; this group has a subgroup of order $\frac{1}{2}(2^4 . 3^4 . 40)$ or $2^3 . 3^4 . 40$, which, as Jordan proved, is *simple*. This subgroup, regarded, as by Jordan, as a group of substitutions of the tritangent planes, contains only *even* substitutions of these. It is this subgroup which is considered here.

A LOCUS WITH 25920 LINEAR SELF-TRANSFORMATIONS

(1) **The fundamental notation.** In a space of four dimensions we may use for homogeneous coordinates the six

$$x_1, \ x_2, \ x_3, \ x_4, \ x_5, \ x_6,$$

connected by the equation $x_1 + x_2 + \ldots + x_6 = 0$. Throughout we shall use ϵ for the cube-root of unity, $\exp(2\pi i/3)$. If l, m, n, p, q, r denote the numbers 1, 2, 3, 4, 5, 6, in any order, the symbol $(lp.mq.nr)$ shall denote the point for which $x_l = x_p = 1$, $x_m = x_q = \epsilon$, $x_n = x_r = \epsilon^2$. The same point is then represented by the symbol $(mq.nr.lp)$, or by the symbol $(nr.lp.mq)$. Occasionally, a couple such as l, p may be spoken of as a *duad*, and the symbol $(lp.mq.nr)$, whose three duads contain all the numbers 1, 2, ..., 6; may be spoken of as a *syntheme* (after Sylvester). Similarly the symbol $[lp.mq.nr]$ shall denote the linear function of the coordinates $x_l + x_p + \epsilon(x_m + x_q) + \epsilon^2(x_n + x_r)$; the prime represented by $[lp.mq.nr] = 0$ is the same as either of those denoted by $[mq.nr.lp] = 0$, or $[nr.lp.mq]$. Further (lp) shall denote the point whose coordinates are $x_l = 1$, $x_p = -1$, with $x_m = x_q = x_n = x_r = 0$, and $[lp]$ shall denote the linear function $x_l - x_p$.

We consider the square scheme of synthemes which, for facility of reference, is printed as frontispiece of the volume, of which the columns are denoted respectively by $\{A\}$, $\{B\}$, $\{C\}$, $\{D\}$, $\{E\}$, $\{F\}$ and the rows by $\{A_0\}$, $\{B_0\}$, $\{C_0\}$, $\{D_0\}$, $\{E_0\}$, $\{F_0\}$. In each column, and in each row, all the fifteen duads of two numbers from 1, 2, 3, 4, 5, 6 occur, each once; the thirty synthemes which occur are all different, but to a syntheme occurring in any row and column there corresponds a syntheme, occurring in the same column and row, differing from the former syntheme by the interchange of the order of the second and third duads. Such a scheme is referred to by Sylvester (*Coll. Papers*, I (1844), p. 92, and *Coll. Papers*, II (1861), p. 265), and may be used in connexion with the theory of the Pascal lines of six points of a conic (Baker, *Principles of Geometry* II (1930), p. 221). But in both these cases the order of the duads in any syntheme is indifferent, while here this order is of the essence of the notation. In this scheme, any duad occurs once in any row, and once in any column; and any two duads that occur once together occur also together in another syntheme, but in reverse order. The scheme thus represents thirty points, each of which can also be characterized by the row and column in which it appears, and denoted by a symbol (PQ_0), or $(Q_0 P)$, where P is one of $A, B, ..., F$, and Q_0 is one of $A_0, B_0, ..., F_0$. For instance $(14.36.25)$ may be denoted by (AB_0). It *will be convenient then to denote* $[14.36.25]$ by $[AB_0]$; *and so in general.*

It may then be verified that in any column, as in any row, the five points represented by the synthemes form a *simplex* in the space of four dimensions, any four of these defining a prime (or space of three dimensions). In fact, the points in the first column, other than the first of these, $(14.36.25)$, or (AB_0), define the prime $[14.25.36] = 0$, or $[A_0 B] = 0$; and the points in the first row, other than the first of these, $(14.25.36)$, or $(A_0 B)$, define the prime $[14.36.25] = 0$, or $[AB_0] = 0$. Or, generally, if P, Q denote two of the letters $A, B, ..., F$, the points of the column $\{P\}$ other than (PQ_0) determine the prime $[P_0 Q] = 0$; and the points of the row $\{Q_0\}$ other than (PQ_0) equally lie in this prime $[P_0 Q] = 0$. (This prime, $[P_0 Q] = 0$, thus contains eight points, and we shall see that it also contains the point $(P Q)$, as well as the three points (lp), (mq), (nr), if $(PQ_0) = (lp.mq.nr)$.) As we have said, in all cases the synthemes $(P_0 Q)$, or $[P_0 Q]$, are obtained respectively from (PQ_0), or $[PQ_0]$, by interchange of the second and third duads of the syntheme.

We shall speak of the simplex, whose angular points are those of the column $\{P\}$, as a *pentahedron* $\{P\}$, and equally of the simplex, whose angular points are those of the row $\{Q_0\}$, as the pentahedron $\{Q_0\}$. Thus the pentahedron $\{A\}$ has, for its angular points, the points $(AB_0), (AC_0), ..., (AF_0)$, and has for prime faces respectively opposite to these the primes $[A_0 B] = 0, [A_0 C] = 0, ..., [A_0 F] = 0$. Similarly the pentahedron $\{B_0\}$ has for its angular points $(AB_0), (CB_0), (DB_0), ..., (FB_0)$, and for opposite prime faces respectively $[A_0 B] = 0, [C_0 B] = 0, ..., [F_0 B] = 0$. We shall also speak of the twelve pentahedra so arising as *Jordan pentahedra*, since the angular points of any one of these essentially occur, associated together, in Jordan's theory of the group of the lines of a cubic surface.

We shall also consider, however, fifteen other Jordan pentahedra. Suppose that the syntheme (PQ_0) is $(lp.mq.nr)$, so that $(P_0 Q)$ is $(lp.nr.mq)$. It can then be verified that the five points $(PQ_0), (P_0 Q), (lp), (mq), (nr)$ form a simplex, with prime faces respectively opposite to these given by $[P_0 Q] = 0, [PQ_0] = 0, [lp] = 0, [mq] = 0, [nr] = 0$. This simplex we speak of as the pentahedron $\{PQ\}$. In addition then to the forty-five points $(PQ_0), (lp)$, we consider, in all, twenty-seven pentahedra, $\{P\}, \{Q_0\}, \{PQ\}$. An angular point of one of these pentahedra will be called the *pole* of the opposite prime face of this pentahedron, this being spoken of as the *polar prime* of the opposite angular point. If $\xi_1, \xi_2, ..., \xi_6$ be the coordinates of one of the forty-five points, and the equation of its polar prime be written

$$u_1 x_1 + u_2 x_2 + ... + u_6 x_6 = 0,$$

where the indefiniteness of $u_1, ..., u_6$ which arises from

$$x_1 + x_2 + ... + x_6 = 0$$

is removed by making the condition $u_1 + u_2 + ... + u_6 = 0$, then, in every case, u_i is the conjugate imaginary of ξ_i, or, say $u_i = \bar{\xi}_i$. If one of the forty-five primes, say the polar of $(\xi_1, \xi_2, ..., \xi_6)$, contain a particular one, $(\eta_1, \eta_2, ..., \eta_6)$, of the forty-five points, then the polar prime of this point contains the pole of the prime in question; for if $\bar{\xi}_1 x_1 + ... + \bar{\xi}_6 x_6 = 0$ contain $(\eta_1, ..., \eta_6)$, then

$$\bar{\xi}_1 \eta_1 + ... + \bar{\xi}_6 \eta_6 = 0,$$

and this involves $\bar{\eta}_1\xi_1 + \dots + \bar{\eta}_6\xi_6 = 0$, where $\bar{\xi}_i$ denotes the conjugate imaginary of ξ_i, etc. Each of the forty-five points is an angular point of three of the pentahedra; for instance (AB_0) is an angular point of the three pentahedra $\{A\}$, $\{B_0\}$, $\{AB\}$; and (14) is an angular point of $\{AB\}$, $\{CD\}$ and $\{EF\}$, since, as the fundamental scheme given above shews, the duad 14 occurs in the synthemes (AB_0) or (A_0B), (CD_0) or (C_0D), (EF_0) or (E_0F); and we have in fact $3.45 = 5.27$. Dually, the opposite prime faces of the three pentahedra which have a common angular point coincide in the polar prime of this point, which thus contains the twelve angular points of these three pentahedra other than their common angular point. For instance, if the angular point be (AB_0), the twelve points (AC_0), (AD_0), (AE_0), (AF_0); (B_0C), (B_0D), (B_0E), (B_0F); (A_0B), (14), (36), (25), all lie in $[A_0\dot{B}] = 0$; or again, if the angular point be (14), the twelve points (AB_0), (A_0B), (36), (25); (CD_0), (C_0D), (23), (56); (EF_0), (E_0F), (26), (35), which can be written down by the fundamental scheme, since $(AB_0) = (14.36.25)$, $(CD_0) = (14.23.56)$, $(EF_0) = (14.26.35)$, all lie in the prime $[14] = 0$. Algebraically, if $(\xi_1, \dots, \xi_6)$, $(\eta_1, \dots, \eta_6)$ be two angular points of the same pentahedron, we have

$$\bar{\xi}_1\eta_1 + \dots + \bar{\xi}_6\eta_6 = 0.$$

Dually, each of the forty-five primes contains three sets of four, of the forty-five points, each set consisting of the angular points of a pentahedron whose other angular point is the pole of the prime. Any one of these forty-five primes may be spoken of as a *Jordan prime*.

(2) **The equation of the Burkhardt primal.** Now consider the locus, in the space of four dimensions in which the coordinates are $x_1, x_2, \dots, x_6$, subject to $x_1 + x_2 + \dots + x_6 = 0$, which is represented by the equation $f = 0$, where

$$f = \Sigma x_l x_m x_n x_p$$

is the sum of the fifteen products of four different coordinates. As has been said, this explicit equation is given by Coble (*loc. cit.*), but such coordinates as $x_1, \ldots, x_6$ were suggested by Burkhardt, and their expression given in terms of the coordinates he employed (*Math. Ann.* xxxviii, p. 205).

Then it is easy to verify that the forty-five points, consisting of the thirty points (PQ_0), and the fifteen points (lp), are all conical nodes of the primal. For the equation $f = 0$ is of the form

$$(x_l + x_m + x_n)\, x_p x_q x_r + (x_p + x_q + x_r)\, x_l x_m x_n$$
$$+ (x_m x_n + x_n x_l + x_l x_m)\,(x_q x_r + x_r x_p + x_p x_q) = 0,$$

where l, m, n, p, q, r denote $1, 2, \ldots, 6$ in any order. And the conditions for a node (in virtue of the relation $x_1 + \ldots + x_6 = 0$) are that all the first derivatives $\partial f/\partial x_i$ should be equal; also

$$\partial f/\partial x_l = x_p x_q x_r + (x_p + x_q + x_r)\, x_m x_n$$
$$+ (x_m + x_n)\,(x_q x_r + x_r x_p + x_p x_q),$$
$$\partial f/\partial x_p = (x_l + x_m + x_n)\, x_q x_r + x_l x_m x_n$$
$$+ (x_m x_n + x_n x_l + x_l x_m)\,(x_q + x_r);$$

thus, at the point $(lp \, . \, mq \, . \, nr)$, for which

$$(x_l, x_m, x_n, x_p, x_q, x_r) = (1, \epsilon, \epsilon^2, 1, \epsilon, \epsilon^2),$$

since

$$x_p x_q x_r = 1, \quad x_p + x_q + x_r = 0, \quad x_q x_r + x_r x_p + x_p x_q = 0,$$
$$x_l + x_m + x_n = 0, \quad x_l x_m x_n = 1, \quad x_m x_n + x_n x_l + x_l x_m = 0,$$

we see that all the six derivatives are equal to 1; and at the point (lp), for which $x_l = 1$, $x_p = -1$, $x_m = x_q = x_n = x_r = 0$, all the six derivatives vanish.

Conversely it is easy to verify, from a more compendious form of the equation of the primal in terms of five homogeneous variables, which occurs below (§ (13)), that the primal has no other nodes than these forty-five.

(3) **Similarity, or equal standing, of the forty-five nodes, and of the twenty-seven pentahedra.** It is clear, from the symmetry of the equation of the primal, that the thirty nodes $(lp \, . \, mq \, . \, nr)$ are entirely similar to one another, and the fifteen nodes (lp) are likewise similar to one another. Likewise that the

twelve pentahedra $\{P\}$, $\{Q_0\}$ are similar to one another, and the fifteen pentahedra $\{PQ\}$ are similar to one another. In fact, *any* two of the nodes, and *any* two of the pentahedra, are similar to one another, notwithstanding the difference of notation. This will appear at once if we put down a linear transformation of the coordinates which leaves the equation of the primal unaltered and changes any chosen one of the fifteen pentahedra $\{PQ\}$, say the pentahedron $\{AB\}$, into one of the twelve pentahedra $\{P\}$, $\{Q_0\}$. Many such transformations are possible; we choose that given by

$$x_1' = x_1 + \epsilon x_2 + \epsilon x_3, \quad -x_4' = x_4 + \epsilon^2 x_5 + \epsilon^2 x_6,$$
$$x_2' = \epsilon x_1 + x_2 + \epsilon x_3, \quad -x_5' = \epsilon^2 x_4 + x_5 + \epsilon^2 x_6,$$
$$x_3' = \epsilon x_1 + \epsilon x_2 + x_3, \quad -x_6' = \epsilon^2 x_4 + \epsilon^2 x_5 + x_6,$$

where, as before $\epsilon = \exp(2\pi i/3)$. These equations, which we shall in future denote by $(x') = \chi(x)$, lead to

$$x_1' + x_2' + \ldots + x_6' = (1 + 2\epsilon)(x_1 + x_2 + \ldots + x_6) = 0,$$

and, if $f(x_1, \ldots, x_6) = 0$ be the equation of the primal, lead to $f(x_1', \ldots, x_6') = 9f(x_1, \ldots, x_6) = 0$, as may be verified without difficulty. The primes of the pentahedron $\{AB\}$ in the new coordinates, say

$$[AB_0]' = 0, \ [A_0B]' = 0, \ [14]' = 0, \ [25]' = 0, \ [36]' = 0,$$

that is

$$x_1' + x_4' + \epsilon(x_3' + x_6') + \epsilon^2(x_2' + x_5') = 0, \ \ldots, \ x_1' - x_4' = 0, \ \ldots,$$

are easily seen to be given by

$$[AB_0]' = (1-\epsilon)[CA_0], \quad [A_0B]' = (1-\epsilon)[CB_0], \quad [14]' = [CD_0],$$
$$[25]' = \epsilon[CE_0], \qquad\qquad [36]' = \epsilon[CF_0].$$

Thus the primes of $\{AB\}$ are changed into the primes of $\{C_0\}$, and consequently the angular points of $\{AB\}$ into the angular points of $\{C_0\}$. For instance, the point (C_0D), or $(14.56.23)$, with coordinates $(x_1, \ldots, x_6) = (1, \epsilon^2, \epsilon^2, 1, \epsilon, \epsilon)$, gives rise to

$$x_1' = -x_4' = 3, \quad x_2' = x_3' = x_5' = x_6' = 0,$$

that is $(14)'$.

This is sufficient for our purpose. In fact the similarity of the twenty-seven pentahedra, and of the forty-five nodes, will be ·continually in evidence in what follows.

(4) **The Jacobian planes of the primal.** If i, j, k be any three of the numbers $1, 2, \ldots, 6$, the equations $x_i/1 = x_j/\epsilon = x_k/\epsilon^2$ represent a plane; and this plane lies entirely on the primal. For, these equations lead to $x_i + x_j + x_k = 0$, $x_j x_k + x_k x_i + x_i x_j = 0$, and, if l, m, n be the three numbers of $1, 2, \ldots, 6$ other than i, j, k, we also have $x_l + x_m + x_n = 0$; so that the equations identically satisfy the equation of the primal. There are twenty sets i, j, k; and the plane $x_i/1 = x_j/\epsilon^2 = x_k/\epsilon$ equally lies on the primal. *There are thus forty planes*, with equations of these forms, which lie on the primal.

It is found on examination that every one of these forty planes contains nine of the forty-five nodes of the primal, these nodes being arranged like the inflexions of a plane cubic curve, so that they lie in threes upon twelve lines, of which four pass through every one of the nine nodes, the joining line of any two of the nodes containing a third node, and the twelve lines form four triangles such that the nine nodes lie in threes upon the three sides of any one of these triangles. As this configuration of nine points was studied by Jacobi, we shall call any one of the forty planes a *Jacobian plane*, to facilitate reference to them.

Consider, for instance, the plane $x_1/1 = x_2/\epsilon = x_3/\epsilon^2$. This evidently contains the collinear points P, Q, R, respectively (56), (64), (45), for every one of which $x_1 = x_2 = x_3 = 0$. Also it contains the points

$$(B_0 C), \text{ or } (15.26.34); \quad (C_0 A), \text{ or } (16.24.35);$$
$$(A_0 B), \text{ or } (14.25.36),$$

with respective coordinates

$$(1, \epsilon, \epsilon^2, \epsilon^2, 1, \epsilon); \quad (1, \epsilon, \epsilon^2, \epsilon, \epsilon^2, 1); \quad (1, \epsilon, \epsilon^2, 1, \epsilon, \epsilon^2),$$

which we denote by L, M, N. These are connected by

$$L + \epsilon^2 M + \epsilon N = 0,$$

so that they are collinear.

§4. THE JACOBIAN PLANES

It also contains the points L', M', N', given respectively by

(EF_0), or $(14.26.35)$; (FD_0), or $(16.25.34)$; (DE_0), or $(15.24.36)$, with coordinates

$$(1, \epsilon, \epsilon^2, 1, \epsilon^2, \epsilon); \quad (1, \epsilon, \epsilon^2, \epsilon^2, \epsilon, 1); \quad (1, \epsilon, \epsilon^2, \epsilon, 1, \epsilon^2),$$

which equally lie on a line, since $L' + \epsilon^2 M' + \epsilon N' = 0$. And the three lines joining any one of P, Q, R to the three points L, M, N, each contain one of the points L', M', N'. The twelve lines each containing three of the points are, in fact,

$$PQR, \quad PMN', \quad PNL', \quad PLM',$$
$$LMN, \quad QNM', \quad QLN', \quad QML',$$
$$L'M'N', \quad RLL', \quad RMM', \quad RNN',$$

of which the three lines in any one of the four columns contain all the nine nodes, and four of the lines pass through any one of the nodes. The existence of the nine lines containing one of the points P, Q, R, one of the points L, M, N, and one of the points L', M', N', is at once seen if we notice such facts as that, for two points $(lp.mq.nr)$, $(lp.mr.nq)$, of which the symbols are obtained from one another by transposition of the numbers q, r, we have the identity

$$(lp.mq.nr) - (lp.mr.nq) = (\epsilon - \epsilon^2)(qr).$$

Moreover, eight such Jacobian planes pass through any one of the forty-five nodes. For through the node $(lp.mq.nr)$ there pass the eight planes $x_i/1 = x_j/\epsilon = x_k/\epsilon^2$, in which i may be l or p, and j may be m or q, and k may be n or r. While, through (for instance) the node (56), there pass the eight planes $x_i/1 = x_j/\epsilon = x_k/\epsilon^2$, $x_i/1 = x_j/\epsilon^2 = x_k/\epsilon$, in which i, j, k are any three of the four numbers $1, 2, 3, 4$.

If, for a moment, we use non-homogeneous coordinates, $X = 0$, $Y = 0$, $Z = 0$, $T = 0$ being four primes which pass through a chosen node, the equation of the primal will be of the form $U_2 + U_3 + U_4 = 0$, where U_i is a homogeneous polynomial of order i in X, Y, Z, T. And, if a plane containing this node lie entirely on the primal, any line in this plane, through this node, will lie entirely on the primal, and so will lie on every one of the cones $U_2 = 0$, $U_3 = 0$, $U_4 = 0$; the plane therefore lies on each of these cones. In particular, then, the eight Jacobian planes through the node lie on the quadric cone $U_2 = 0$, and are thus identified as the intersection of this cone with the primal. This cone we call the *asymptotic cone* at the node. A quadric cone such as $U_2 = 0$, in space of four dimensions, contains two systems of planes, each consisting of ∞^1 planes, with the property that two planes of the same system have in common only the point-vertex of the cone, while two planes of different systems meet in a line passing through this vertex; in fact the cone meets a threefold space in a quadric surface whose generating lines are projected from the vertex by the planes of the cone. It can at once be verified that the eight Jacobian planes through any node of the primal consist of four planes of the asymptotic cone of one system, together with four planes of the other system; so that there are sixteen lines through the node, lying on the primal, through each of which two Jacobian planes pass. For instance, consider the Jacobian planes through the node (14.25.36): the planes $x_1/1 = x_2/\epsilon = x_3/\epsilon^2$ and $x_4/1 = x_5/\epsilon = x_3/\epsilon^2$ meet only in the node, whose coordinates are $(1, \epsilon, \epsilon^2, 1, \epsilon, \epsilon^2)$, but the planes

$$x_1/1 = x_2/\epsilon = x_3/\epsilon^2 \text{ and } x_4/1 = x_2/\epsilon = x_3/\epsilon^2$$

meet in the line $x_1/1 = x_4/1 = x_2/\epsilon = x_3/\epsilon^2$. Or again, considering the planes through the node (56), the planes $x_1/1 = x_2/\epsilon = x_3/\epsilon^2$ and $x_4/1 = x_2/\epsilon^2 = x_3/\epsilon$ meet only in the node, for which $x_1 = x_2 = x_3 = x_4 = 0$, while the planes $x_1/1 = x_2/\epsilon = x_3/\epsilon^2$ and $x_4/1 = x_2/\epsilon = x_3/\epsilon^2$ meet in the line $x_1/1 = x_4/1 = x_2/\epsilon = x_3/\epsilon^2$. The same fact will appear later from another form of the equation of the primal. Also it will appear immediately that the four lines in which a Jacobian plane, α, through a node, is met by the four planes of the opposite system of the asymptotic cone at that node, are the lines, in the plane α, through the node, the vertex of the cone, which each contain two other nodes of the plane α.

The Jacobian planes considered have each twelve lines containing three nodes apiece. More generally, it is true that, if an *arbitrary* plane be drawn through any line which contains three nodes of the primal, its residual intersection with the primal is a cubic curve having an inflexion at each of the three nodes. For instance, consider the three collinear nodes (56), (64), (45), at all of which $x_1 = x_2 = x_3 = 0$. Let a, b, c be arbitrary numbers and $p_1 = a+b+c$, $p_2 = bc+ca+ab$, $p_3 = abc$. The equations of the general plane containing these three nodes are $x_1/a = x_2/b = x_3/c$, and for the intersection with the primal each of these fractions is equal to $-(x_4+x_5+x_6)/p_1$. Thus, from the equation of the primal, we find that the residual intersection of the plane lies on the cubic curve obtainable from

$$p_1^3 x_4 x_5 x_6 + p_3 (x_4+x_5+x_6)^3$$
$$-p_1 p_2 (x_4+x_5+x_6)(x_5 x_6+x_6 x_4+x_4 x_5) = 0;$$

interpreting x_4, x_5, x_6 as coordinates in a plane, this curve has an inflexion at each of the three points $(0, 1, -1)$, $(-1, 0, 1)$, $(1, -1, 0)$, the inflexional tangent at the first of these having the equation $p_1^2 x_4 - p_2 (x_4+x_5+x_6) = 0$; and so on.

(5) **The x-lines of the primal.** It may easily be verified that through any line containing three nodes in a Jacobian plane there passes also another Jacobian plane. For instance, recurring to the enumeration of the nodes in the plane $x_1/1 = x_2/\epsilon = x_3/\epsilon^2$ given in the preceding §(4), the nodes P, Q, R also lie on the Jacobian plane $x_1/1 = x_2/\epsilon^2 = x_3/\epsilon$, the nodes L, M, N also lie on the Jacobian plane $x_4/1 = x_5/\epsilon = x_6/\epsilon^2$, and the nodes L', M', N' on the plane $x_4/1 = x_5/\epsilon^2 = x_6/\epsilon$. Likewise the nodes P, M, N', or, respectively, (56), $(1, \epsilon, \epsilon^2, \epsilon, \epsilon^2, 1)$, $(1, \epsilon, \epsilon^2, \epsilon, 1, \epsilon^2)$, lie on the plane $x_1/1 = x_3/\epsilon^2 = x_4/\epsilon$; the nodes P, N, L' or, respectively, (56), $(1, \epsilon, \epsilon^2, 1, \epsilon, \epsilon^2)$, $(1, \epsilon, \epsilon^2, 1, \epsilon^2, \epsilon)$, lie on the plane $x_2/\epsilon = x_3/\epsilon^2 = x_4/1$; and the nodes P, L, M', or, respectively, (56), $(1, \epsilon, \epsilon^2, \epsilon^2, 1, \epsilon)$, $(1, \epsilon, \epsilon^2, \epsilon^2, \epsilon, 1)$, lie on the plane $x_1/1 = x_2/\epsilon = x_4/\epsilon^2$; and so on—and, in virtue of the symmetry of the equation of the primal, this

formulation is quite general. Thus, taking the four Jacobian planes of one system lying on the asymptotic cone at a particular node of the primal, and, in each of these planes, the four lines through the node which each contain two other nodes, we obtain sixteen lines through the node, each the intersection of two Jacobian planes; the aggregate of the Jacobian planes thus arising must then coincide with the eight Jacobian planes passing through the node, which form the complete intersection of the primal with the asymptotic cone considered. The sixteen lines which are the intersections of these planes are therefore such that each contains three nodes. From the property of a quadric cone in space of four dimensions the two Jacobian planes which intersect in any one of these lines determine the tangent prime of the asymptotic cone at any ordinary point of this line; it will appear later that this prime also touches the Burkhardt primal at any ordinary point of this line. In this way then we obtain $\frac{1}{2}$ 40.12, or 15.16, or 240 lines, lying on the primal, of which each contains three nodes, this number being also $\frac{1}{3}$ 45.16. Conversely, any line on the primal which contains three nodes can be shewn to lie on two Jacobian planes (cf. § (4)). Thus 240 is the total number of such lines. *Such lines on the primal, containing three nodes, we shall call κ-lines.*

And, as there are sixteen κ-lines through each of the forty-five nodes, so there are in fact sixteen κ-lines in each of the forty-five polar primes of these nodes, or, as we say, in each of the *Jordan primes* which are the prime faces of the twenty-seven Jordan pentahedra. We have seen that there are, in each such prime, three sets of four points, each such set consisting of four angular points of one of the three pentahedra which have the pole of this prime as common angular point. These sets form three tetrahedra in this prime. It is the case that the four lines joining any one of the angular points, of one of these tetrahedra, to all the angular points of another tetrahedron, pass each through an angular point of the third tetrahedron. The sixteen lines so obtained each contain three nodes, the angular points of three tetrahedra. We may verify this statement in two representative cases:—

(a) The prime $[A_0 B] = 0$, polar of the node $(A B_0)$, contains the other angular points of the three pentahedra $\{AB\}$, $\{A\}$, $\{B_0\}$, which have $(A B_0)$ as a common angular point, as we have remarked. The prime thus contains the three tetrahedra whose angular points are

$$(A_0 B), (14), (36), (25); \quad (AC_0), (AD_0), (AE_0), (AF_0);$$
$$(B_0 C), (B_0 D), (B_0 E), (B_0 F),$$

which we may denote respectively by

$$P, Q, R, S; \quad L, M, N, T; \quad L', M', N', T';$$

it is then found that these lie in threes on the sixteen lines

$$(PLL', PMM', PNN', PTT'); \quad (QLM', QML', QNT', QTN');$$
$$(RLT'', RMN', RNM', RTL'); \quad (SLN', SMT'', SNL', STM').$$

The verification of this, with help of the scheme in § (1), depends on the two facts that if U, V, W be any three of A, B, C, D, E, F, then the three nodes $(U V_0)$, $(V W_0)$, $(W U_0)$ lie on a line (for instance we find $(A B_0) + \epsilon(B C_0) + \epsilon^2(C A_0) = 0$); and, if $(lp.mq.nr)$, $(lp.mr.nq)$ be the symbols of two nodes, differing only by the transposition of the numbers q, r, then these two nodes lie on a line which contains the node (qr).

(b) The prime $[14] = 0$, polar of the node (14), contains the other angular points of the pentahedra $\{AB\}$, $\{CD\}$, $\{EF\}$; these form the three tetrahedra whose angular points are

$$(AB_0), (A_0 B), (36), (25); \quad (CD_0), (C_0 D), (23), (56);$$
$$(EF_0), (E_0 F), (26), (35),$$

namely
$$(14.36.25), (14.25.36), (36), (25);$$
$$(14.23.56), (14.56.23), (23), (56);$$
$$(14.26.35), (14.35.26), (26), (35),$$

or say

$$L, M, X, Y; \quad L', M', X', Y'; \quad L'', M'', X'', Y'';$$

it is then clear that $L'L''X$, $L'M''Y$, $M'L''Y$, $M'M''X$, are lines, and two other sets of four lines are similarly found by combining

the third and first tetrahedra and the first and second tetrahedra;
while evidently $XX'X''$, $XY'Y''$, $YX'Y''$, $YY'X''$ are also lines.
Three tetrahedra related as here are said to be a desmic
system. They are particular cases of quartic surfaces whose
equation, in space of three dimensions, with coordinates x, y, z, t,
can be taken in the form

$$a(y^2z^2 + x^2t^2) + b(z^2x^2 + y^2t^2) + c(x^2y^2 + z^2t^2) = 0,$$

with $a + b + c = 0$ (cf. Note 1 at the end of this volume). The section
of the Burkhardt primal by one of the Jordan primes is a particular quartic surface of this form. An account of desmic surfaces
is given in Jessop, *Quartic Surfaces*, Cambridge (1916), Chap. II.

It follows from what we have proved, if we consider a particular
Jordan pentahedron, say $\{A\}$, that, in each of the four primes of
the pentahedron which meet in a particular angular point, say
(AB_0), there are four κ-lines passing through that angular point.
This gives the sixteen κ-lines which we have already shewn to
pass through the node (AB_0). But there are two other pentahedra
with the same angular point, namely $\{B_0\}$ and $\{AB\}$, and each of
these equally gives rise to sixteen κ-lines through (AB_0). These
then coincide with the former set. In fact, through each of the
sixteen κ-lines from (AB_0), there passes a prime face of each of the
pentahedra $\{A\}$, $\{B_0\}$, $\{AB\}$. These three prime faces do not
determine the line; they are, as we shall see immediately, the
polar primes of three nodes lying on a line, and meet in a plane.
We can verify the identity

$$[A_0C][A_0D][A_0E][A_0F] - [BC_0][BD_0][BE_0][BF_0]$$
$$= 3(\epsilon^2 - \epsilon)[14][25][36][AB_0],$$

connecting the twelve primes through (AB_0) which are the faces
respectively of the pentahedra $\{A\}$, $\{B_0\}$, $\{AB\}$ (the algebra is
straightforward if we express the left side in terms of the six
quantities $a = x_1 + \epsilon x_4$, $a_0 = x_1 + \epsilon^2 x_4$, $b = x_2 + \epsilon x_5$, $b_0 = x_2 + \epsilon^2 x_5$,
$c = x_3 + \epsilon x_6$, $c_0 = x_3 + \epsilon^2 x_6$). This shews that the plane of intersection of any one of the first four primes with any one of the

second four primes lies on one of the four primes occurring on the right side. Thus there are sixteen planes through the node (AB_0), each lying in three primes, taken respectively from one of the three products of four. Also, for instance, the four κ-lines through (AB_0), in the face $[A_0C] = 0$ of the pentahedron $\{A\}$, containing respectively, besides (AB_0), the pairs of nodes

$$(BC_0), (CA_0); \quad (C_0D), (35); \quad (C_0E), (16); \quad (C_0F), (24),$$

lie, respectively, in planes given by the pairs of primes

$$[BC_0] = 0, [AB_0] = 0; \quad [BD_0] = 0, [14] = 0;$$

$$[BE_0] = 0, [25] = 0; \quad [BF_0] = 0, [36] = 0,$$

as well as in the prime $[A_0C] = 0$; and further these four lines lie, respectively, in the pairs of Jacobian planes

$$\left.\begin{array}{l} x_1/1 = x_2/\epsilon^2 = x_3/\epsilon \\ x_4/1 = x_5/\epsilon^2 = x_6/\epsilon \end{array}\right\}; \quad \left.\begin{array}{l} x_1/1 = x_2/\epsilon^2 = x_6/\epsilon \\ x_2/1 = x_4/\epsilon = x_6/\epsilon^2 \end{array}\right\};$$

$$\left.\begin{array}{l} x_2/1 = x_3/\epsilon^2 = x_4/\epsilon \\ x_3/1 = x_4/\epsilon^2 = x_5/\epsilon \end{array}\right\}; \quad \left.\begin{array}{l} x_1/1 = x_3/\epsilon = x_5/\epsilon^2 \\ x_1/1 = x_5/\epsilon^2 = x_6/\epsilon \end{array}\right\}.$$

These statements appear at once by reference to the fundamental scheme given in the frontispiece.

The κ-lines thus arise as twelve in every Jacobian plane, sixteen in every Jordan prime, and sixteen through every node. As the lines are of importance for the theory of the transformations of the Burkhardt primal into itself by projections, which arises below, we enumerate the totality of the κ-lines also as follows. They may be regarded as of three sorts:

(a) Those containing three of the fifteen nodes (ij), arising from identities such as $(ij) + (jk) + (ki) = 0$. These are twenty in number, and form a configuration, with these fifteen nodes, familiar in the discussion of Desargues' theorem for triangles in perspective in space of three dimensions, but existing in general form in fourfold space. This configuration is described by the scheme

$$15(.4, 6, 4)20(3.3, 3)15(6, 4.2)6(10, 10, 5.),$$

which means that each of the fifteen points (ij) lies on four of the
κ-lines considered, and also in six planes, and in four solids; also
that each of the twenty κ-lines contains three of the fifteen points,
and lies in three planes, and in three solids; that there are fifteen
planes each containing six of the nodes, four of the κ-lines, and
lying in two solids; while there are six solids in the figure, each
containing ten of the twenty nodes, ten of the κ-lines, and five of
the planes.

(b) Through each of the fifteen nodes (ij), besides the four
κ-lines enumerated under (a), there pass twelve other κ-lines,
each containing two of the thirty nodes (PQ_0). In fact, among
these thirty nodes, omitting six whose symbol contains the duad
(ij), the other twenty-four lie in pairs on κ-lines through (ij).
Conversely, the lines joining any one of the thirty nodes (PQ_0),
say $(14.25.36)$, to the twelve nodes (ij) other than (14), (25), (36),
each contain another node of the thirty. Thus there are 15.12
κ-lines each containing two of the thirty nodes (PQ_0) and one of
the fifteen nodes (ij). The rule, as we have already remarked, is
that, if the syntheme symbols of two of the thirty nodes are
interchanged by transposition of the numbers i, j, then their join
contains the node (ij).

(c) But, besides the twelve κ-lines through one of the thirty
nodes (PQ_0) which are enumerated under (b), there are four
κ-lines through each of these thirty nodes which each contain
two other of these thirty nodes. For instance, through (AB_0)
there pass the four κ-lines containing (BP_0), (PA_0), where P is
any one of C, D, E, F; for we have (e.g.) the identity

$$(AB_0) + \epsilon(BC_0) + \epsilon^2(CA_0) = 0.$$

We thus obtain $\frac{1}{3}30.4$ or forty κ-lines, each containing three
nodes, such as (PQ_0), (QR_0), (RP_0).

In all, the number of κ-lines thus enumerated under (a), (b), (c),
is $20 + 15.12 + 40$, or 240, which is the complete number.

The forty-five nodes can be joined in pairs in $\frac{1}{2}45.44$ ways, and
there are 10.27 joining lines each of a pair of angular points of

the same pentahedron (if we assume that no two pentahedra have two nodes in common). There remain then $990-270$, or 720, joining lines of a pair of nodes which do not belong to the same pentahedron; we infer then that each of these latter joins contains a third node, so giving rise to one of the $\frac{1}{3} 720$, or 240, existing κ-lines. In other words, *the join of any two of the forty-five nodes is either an edge-line of one of the twenty-seven pentahedra, or is a κ-line, containing another node.*

Now let M, N be two nodes of a κ-line, being therefore angular points of different pentahedra. There are three pentahedra of which M is an angular point, and three pentahedra of which N is an angular point. On examination it is found that every one of the three pentahedra having M as common angular point has a single angular point common with a particular one of the three pentahedra having N as common angular point, and the three new angular points thus arising lie in a line. Each of these three angular points belongs thus to two pentahedra (one from M and one from N), and each of these points is an angular point of a third pentahedron, so that there arise three more pentahedra; these three additional pentahedra have in fact a common angular point, the third node lying on the κ-line MN. Whence, any κ-line LMN gives rise to another κ-line, say $L'M'N'$; and, from this, the original κ-line can be conversely derived by the same construction. From this it follows that, if μ, ν be the polar primes respectively of M and N, then, of the twelve nodes lying in μ, there are three, say L', M', N', which coincide with three of the twelve nodes lying in the prime ν, these being in line; and an enumeration of the two sets of twelve nodes, in μ and ν, immediately shews what L', M', N' are. The polar primes of L, M, N meet in a *plane*, in which the line $L'M'N'$ lies; likewise L, M, N lie in the plane common to the polar primes of L', M', N'.

We shall speak of two κ-lines which are associated in this way as *polar κ-lines*.

Now consider a Jordan pentahedron Ω, of which L is an angular point, the κ-line LMN lying in a face λ', of Ω, which is opposite to another angular point, L', of Ω. Then the polar primes of the two nodes M, N, in λ', necessarily pass through the pole L' of λ', so that L' is one of the nodes of the κ-line polar to LMN. Next, let Ω', Ω'' be the other two pentahedra having L as angular point; both Ω' and Ω'' have prime faces, through L, which contain LMN, as we have seen, and the poles of these faces lie in the prime λ. The polar primes of L, M, N thus pass through these two poles, say M', N', as well as through the node L'. The polar line $L'M'N'$, of the line LMN, is thus one of the lines, in the prime λ, which contains an angular point of each of the three tetrahedra formed by the nodes in λ, these being angular points of Ω, Ω', Ω''. The other three lines of this character, through L', in the prime λ, arise similarly from the other three κ-lines, in the prime λ', through L. By considering the other three prime faces of Ω through L we similarly obtain twelve other κ-lines of the prime λ. And we may in the same way consider every node of Ω, and obtain, corresponding to every κ-line through this, a polar κ-line in the polar prime of this node.

The complete process uses, besides Ω, two pentahedra for each node of Ω; in all eleven pentahedra. There remain, then, from the complete tale of twenty-seven pentahedra, sixteen others. Now, each face of Ω contains eight nodes, other than angular points of Ω, and each such node, defined as an angular point of a pentahedron having one angular point common with Ω, is an angular point of two other pentahedra, neither of which has an angular point common with Ω. We infer, therefore, that there are sixteen pentahedra, of which no one has an angular point common with a specified pentahedron Ω, and that these sixteen pentahedra have their angular points on the faces of Ω, namely at the angular points, eight for each face, of pentahedra having a node common with Ω, and that each of these sixteen pentahedra has its angular points in the five faces of Ω, one in each, but not at an angular point of Ω. The prime faces of any one of these sixteen pentahedra, therefore, each contain one of the five angular points of Ω. If two pentahedra which have no common angular point be spoken of as *skew* to one another, we may thus say that *there are sixteen pentahedra which are skew to any given one, and that any two skew pentahedra are both inscribed and circumscribed to one another, the angular points of either lying severally on the prime faces of the other.*

Two anticipatory remarks may be made in regard to the preceding argument: (*a*) If L, M, N be a κ-line in the face λ' of a pentahedron Ω, through the angular point L, the harmonic conjugate of L in regard to M and N lies on the face λ polar to L; (*b*) Considering the asymptotic cones of the Burkhardt primal at L and L', four primes can be drawn through the line LL', each to touch both these cones along lines, one pair of such lines of contact being the polar κ-lines LMN, $L'M'N'$ in the faces λ' and λ, the other pairs being the other corresponding κ-lines through L and L' in these faces. Both these remarks are established below.

As illustration of the derivation of the polar κ-line of a given κ-line, we may consider the κ-line, in the face $[A_0 C]$ of the pentahedron $\{A\}$, which contains the angular point (AB_0) and also contains the nodes (BC_0), (CA_0). The polar primes of these three nodes each contain twelve nodes, which are, respectively, for the three nodes in turn,

$$(AC_0),\ (AD_0),\ (AE_0),\ (AF_0),\ \text{in } \{A\};$$
$$(B_0 C),\ (B_0 D),\ (B_0 E),\ (B_0 F),\ \text{in } \{B_0\};$$
$$(A_0 B),\ (14),\ (36),\ (25),\ \text{in } \{AB\};$$
$$(BA_0),\ (BD_0),\ (BE_0),\ (BF_0),\ \text{in } \{B\};$$
$$(C_0 A),\ (C_0 D),\ (C_0 E),\ (C_0 F),\ \text{in } \{C_0\};$$
$$(B_0 C),\ (15),\ (34),\ (26),\ \text{in } \{BC\};$$
$$(CB_0),\ (CD_0),\ (CE_0),\ (CF_0),\ \text{in } \{C\};$$
$$(A_0 B),\ (A_0 D),\ (A_0 E),\ (A_0 F),\ \text{in } \{A_0\};$$
$$(C_0 A),\ (16),\ (35),\ (24),\ \text{in } \{CA\},$$

the nodes in $\{AB\}$, $\{BC\}$, $\{CA\}$ being read off from the fundamental scheme of § (1). The first two sets here put down have in

§5. THE κ-LINES OF THE PRIMAL 19

common just the three nodes (AC_0), (CB_0), (BA_0), and these belong to those in the third prime, which contains the plane common to the first two primes. More generally, the polar κ-line of that containing the nodes (PQ_0), (QR_0), (RP_0) is that containing the nodes (P_0Q), (Q_0R), (R_0P).

Or again, consider the κ-line which is the polar of the κ-line containing the nodes (23), (31), (12). As we see from the fundamental scheme given in §(1), the node (23) is the common angular point of the three pentahedra $\{AF\}$, $\{BE\}$, $\{CD\}$; the node (31) is the common angular point of the three pentahedra $\{AD\}$, $\{BF\}$, $\{CE\}$; and the node (12) is the common angular point of the three pentahedra $\{AE\}$, $\{BD\}$, $\{CF\}$. Also the node (56) is common to $\{CD\}$, $\{BF\}$, $\{AE\}$; the node (64) is common to $\{AF\}$, $\{CE\}$, $\{BD\}$; and the node (45) is common to $\{BE\}$, $\{AD\}$, $\{CF\}$. Thus the polar line in question is that containing (56), (64), (45). In general, if l, m, n, p, q, r be the numbers 1, 2, ..., 6, in any order, the two κ-lines (mn), (nl), (lm) and (qr), (rp), (pq) are polars.

Finally, consider such a κ-line as that containing (AB_0), (CD_0), (26), of which the two former have the symbols (14.36.25), (14.23.56), which are interchanged by the transposition of the numbers 2 and 6, there existing the identity

$$(AB_0) - (CD_0) = (\epsilon^2 - \epsilon)(26).$$

To obtain the polar κ-line, we consider the nodes (AD_0), (CB_0), obtained by transposition of B_0 and D_0 in the first κ-line, which have the symbols (13.45.26), (15.26.34); these symbols are interchanged by transposition of the numbers 1 and 4, and there exists the identity $(AD_0) - \epsilon(CB_0) = (1-\epsilon)(14)$. Thus $(A\dot{D}_0)$, (CB_0), (14) belong to a κ-line, and it is easy to see that this is the polar κ-line in question. Here, the duad 26 occurs in the symbols of both (AD_0) and (CB_0), and the duad 14 occurs in the symbols of both (AB_0) and (CD_0). In general, if the symbols of (PQ_0), (RS_0) be interchanged by the transposition of two of the numbers 1, 2, ..., 6, then these two nodes belong to a κ-line, whose polar κ-line joins the nodes (PS_0), (RQ_0).

20   §6. THE PRIMAL IS RATIONAL

(6) **The Burkhardt primal is rational.** In a Jacobian
plane ϖ, take three nodes P, Q, R, which are not in line. Through
each of the κ-lines forming the sides of the triangle PQR, there
passes then another Jacobian plane, as we have said. We easily
shew that these planes α, β, γ, respectively through QR, RP, PQ,
have no intersections besides the points P, Q, R. For, the planes
β, γ lie on the asymptotic cone of the primal at P, and both meet
the plane ϖ, which also lies on this cone, in a line; they are then
of opposite systems to ϖ in the planes of this cone, and therefore
of the same system to one another, and so only meet in P. Likewise for γ, α and for α, β. We may say then that every two of
these planes are skew to one another.

But, in space of four dimensions, a single line can be drawn
through an arbitrary point H, to meet each of three skew planes
α, β, γ, in a point, this being the line common to the three primes
(H, α), (H, β), (H, γ). Taking H on the Burkhardt primal, this
line will meet an arbitrary threefold space Π in a point K, which
then, the planes α, β, γ being given, is determined by H, while,
conversely, there is a definite line through an arbitrary point K of
the space Π, meeting the planes α, β, γ each in a point, and, as
α, β, γ lie on the primal, which is of order 4, this line through K
will meet the primal in a further point H, equally determined by
K. There is thus a definite point to point correspondence between
the primal and the space Π, evidently expressible by rational
equations in both senses. We shall obtain these equations
explicitly for one choice of the planes α, β, γ, in a later section
($\S(14)$).

There are three lines in which the solid Π is met by the planes
α, β, γ, which as α, β, γ lie on the primal, are also on the primal.
Evidently these lines are exceptional in the transformation.

We can shew that there are 72.40 sets of three planes, such as
α, β, γ, lying on the primal, which are mutually skew in pairs.
For first, in any given Jacobian plane, there are $\frac{1}{6} 9.8.7 - 12$, or
seventy-two triangles whose angular points are nodes. Taking all
Jacobian planes we can thus obtain 72.40 such sets of three

planes, no two of which coincide, since the plane containing the three nodes where the planes meet in pairs, coincides in each generation with the Jacobian plane from which the set is generated. Conversely, if α, β, γ be any three mutually skew Jacobian planes lying on the primal, the points $P = (\beta, \gamma)$, $Q = (\gamma, \alpha)$, $R = (\alpha, \beta)$, must be nodes of the primal, since there are lines through each point, lying on the primal, which generate two planes. And each line such as QR, joining two nodes of a Jacobian plane, must be a κ-line. The second Jacobian plane α', through QR must then meet β and γ in lines, as we see by considering the asymptotic cones at Q and R. These two lines, in the plane α', must then meet, and can only meet in the single common point of β and γ, which is P. Thus α' is the plane PQR, which, coinciding with α', is a Jacobian plane. There is then no other way of obtaining three such skew planes as α, β, γ than that followed above. The total number 72.40 of such sets of three planes, is obtainable also in the form $\frac{1}{3}$ 45.8.24, by considering the eight Jacobian planes through every node such as P, and, therein, the triangles such as PQR.

(7) **The particular character of the forty-five nodes, and the linear transformation of the primal into itself by projection from the nodes.** It can be shewn that, if the equation of the Burkhardt primal be expressed in terms of four non-homogeneous coordinates X, Y, Z, T, vanishing at a node, so that its equation takes the form $U_2 + U_3 + U_4 = 0$, in which U_i is homogeneous of order i in X, Y, Z, T, then *the quadric cone $U_2 = 0$ forms part of the cubic cone $U_3 = 0$*; so that the equation is really of the form $U_2(1 + U_1) + U_4 = 0$, where U_1 is linear in X, Y, Z, T. Thus a line through the node which lies on the asymptotic cone $U_2 = 0$, and does not lie entirely on the primal meets the primal only at the node. But a more interesting consequence is that, if a line through the node O meets the primal again in P and P', and H be the harmonic conjugate $(P, P')/O$, of O in regard to P and P', then the locus of H, for all

lines through O, is a prime. The equation of this prime is, in fact, $2 + U_1 = 0$. In terms of five homogeneous variables x, y, z, t, u, if the equation of the primal be denoted by $F_x^4 = 0$, and $(\xi, \eta, \zeta, \tau, \omega)$ be the coordinates of a node, the statement is that the cubic polar $F_x^3 F_\xi$ contains the quadratic polar $F_x^2 F_\xi^2$ as a factor. To express formally the results of this, let the line joining the node $(\xi, \eta, \zeta, \tau, \omega)$ to another point (x, y, z, t, u), lying on the primal, meet the primal again in the point (x', y', z', t', u'); then, substituting for x', y', z', t', u', in the equation of the primal,

$$x' = x + \lambda \xi, \quad \zeta' = y + \lambda \eta, \quad z' = z + \lambda \zeta, \quad t' = t + \lambda \tau, \quad u' = u + \lambda \omega,$$

we have

$$F_x^4 + 4\lambda F_x^3 F_\xi + 6\lambda^2 F_x^2 F_\xi^2 + 4\lambda^3 F_x F_\xi^3 + \lambda^4 F_\xi^4 = 0,$$

wherein $F_x^4 = 0$, $F_\xi^4 = 0$, and $F_x F_\xi^3 = 0$ because $(\xi, ..., \omega)$ is a node. Thus we obtain

$$2F_x^3 F_\xi + 3\lambda F_x^2 F_\xi^2 = 0.$$

Now assume the property stated above, that $F_x^3 F_\xi / F_x^2 F_\xi^2$ is (identically in regard to $x, y, ..., u$) equal to a linear function of $x, y, ..., u$, because of the character of the node, and so put

$$2F_x^3 F_\xi / F_x^2 F_\xi^2 = v_1 x + v_2 y + ... + v_5 u, = v_x, \text{ say;}$$

then we have

$$x' = x - \tfrac{1}{3}\xi . v_x, \quad y' = y - \tfrac{1}{3}\eta . v_x, \quad ..., \quad u' = u - \tfrac{1}{3}\omega . v_x,$$

whereby $x', y', ..., u'$ are expressed as linear functions of $x, y, ..., u$. In terms of the node O, and the points

$$P = (x, y, ..., u), \quad P' = (x', y', ..., u'),$$

this result is expressible in the form

$$(P') - (P) = -\tfrac{1}{3}v_x(O),$$

and for the harmonic point H, of O, in regard to Γ and Γ', we may thus take $(H) = \tfrac{1}{2}[(P') + (P)]$, or $(H) = (P) - \tfrac{1}{6}v_x(O)$. If then $(h_1, h_2, ..., h_5)$ be the coordinates of H, we have

$$v_h = v_x - \tfrac{1}{6}v_x v_\xi = v_x(1 - \tfrac{1}{6}v_\xi).$$

Now, in the identity by which $v_1, \ldots, v_5$ were defined, put $x = \xi + \theta_1$, $y = \eta + \theta_2$, ..., $u = \omega + \theta_5$; thence,

$$v_x = \frac{2(F_\xi + F_\theta)^3 F_\xi}{(F_\xi + F_\theta)^2 F_\xi^2} = \frac{6F_\xi^2 F_\theta^2 + 2F_\theta^3 F_\xi}{F_\theta^2 F_\xi^2} = 6 + \frac{2F_\theta^3 F_\xi}{F_\theta^2 F_\xi^2}$$

$$= 6 + v_1 \theta_1 + v_2 \theta_2 + \ldots + v_5 \theta_5,$$

so that, if we take $\theta_1 = \theta_2 = \ldots = \theta_5 = 0$, we infer that $v_\xi = 6$. Hence we have $v_h = 0$. In other words, as P varies, the harmonic point H describes the prime $v_x = 0$.

The most direct way to prove the property of the nodes of the primal which we have enunciated is to express the equation of the primal in terms of the primes which are the faces of a Jordan pentahedron, as will be done below (§(15)). But we can avoid this, and dispense with the general formulation just given, by examining the cases of two representative nodes, with the equation $\Sigma x_i x_j x_k x_l = 0$ used above.

One such node is (14), or $(1, 0, 0, -1, 0, 0)$. We substitute then

$$x_1' = x_1 + \lambda, \quad x_2' = x_2, \quad x_3' = x_3, \quad x_4' = x_4 - \lambda, \quad x_5' = x_5, \quad x_6' = x_6,$$

in the equation

$$(x_1 + x_2 + x_3) x_4 x_5 x_6 + (x_4 + x_5 + x_6) x_1 x_2 x_3$$
$$+ (x_2 x_3 + x_3 x_1 + x_1 x_2)(x_5 x_6 + x_6 x_4 + x_4 x_5) = 0,$$

and so obtain, as $(x_1, \ldots, x_6)$ and $(x_1', \ldots, x_6')$ lie on the primal, the result

$$(x_4 - x_1 - \lambda) U = 0, \text{ with } U = x_5 x_6 + x_2 x_3 + (x_2 + x_3)(x_5 + x_6).$$

Thus, assuming U not to vanish, in which case the line joining $(x_1, \ldots, x_6)$, $(x_1', \ldots, x_6')$ would lie entirely on the primal, we have $\lambda = x_4 - x_1$. In other words, $(x_1', \ldots, x_6')$ is obtained from $(x_1, \ldots, x_6)$ by interchange of x_1 and x_4. And the harmonic point of the node, in regard to $P = (x_1, \ldots, x_6)$ and $P' = (x_1', \ldots, x_6')$, which is $\frac{1}{2}[(P) + (P')]$ or

$$(P) + \tfrac{1}{2}\lambda(1, 0, 0, -1, 0, 0), \text{ or } [\tfrac{1}{2}(x_4 + x_1), x_2, x_3, \tfrac{1}{2}(x_4 + x_1), x_5, x_6],$$

describes the prime $[14] = 0$, namely the polar prime of the node (14).

Another representative node is $(A_0 B)$, or $(14.25.36)$, with coordinates $(1, \epsilon, \epsilon^2, 1, \epsilon, \epsilon^2)$. If then we substitute in the equation of the primal the values

$$x_1' = x_1 + \lambda, \quad x_2' = x_2 + \epsilon\lambda, \quad x_3' = x_3 + \epsilon^2\lambda,$$
$$x_4' = x_4 + \lambda, \quad x_5' = x_5 + \epsilon\lambda, \quad x_6' = x_6 + \epsilon^2\lambda,$$

using the facts that $(x_1, \ldots, x_6)$ lies on the primal, and

$$x_1 + \ldots + x_6 = 0,$$

we find, with $U = x_1 x_4 + \epsilon^2 x_2 x_5 + \epsilon x_3 x_6$, that

$$[x_1 + x_4 + \epsilon^2(x_2 + x_5) + \epsilon(x_3 + x_6) + 3\lambda] U = 0,$$

so that, unless $U = 0$, in which case the joining line of $(x_1, \ldots, x_6)$ and $(x_1', \ldots, x_6')$ would lie entirely on the primal, we have

$$\lambda = -\tfrac{1}{3}[A B_0],$$

and the equations of transformation are

$$x_1' = x_1 - \tfrac{1}{3}[A B_0], \quad x_2' = x_2 - \frac{\epsilon}{3}[A B_0], \quad x_3' = x_3 - \frac{\epsilon^2}{3}[A B_0],$$

$$x_4' = x_4 - \tfrac{1}{3}[A B_0], \quad x_5' = x_5 - \frac{\epsilon}{3}[A B_0], \quad x_6' = x_6 - \frac{\epsilon^2}{3}[A B_0].$$

The harmonic point of the node in regard to $(x_1, \ldots, x_6)$ and $(x_1', \ldots, x_6')$, say $(X_1, \ldots, X_6)$, being

$$X_i = x_i - \tfrac{1}{6}[A B_0], \quad X_j = x_j - \frac{\epsilon}{6}[A B_0], \quad X_k = x_k - \frac{\epsilon^2}{6}[A B_0],$$

$(i = 1, 4; j = 2, 5; k = 3, 6)$, is such that

$$X_1 + X_4 + \epsilon(X_3 + X_6) + \epsilon^2(X_2 + X_5)$$
$$= x_1 + x_4 + \epsilon(x_3 + x_6) + \epsilon^2(x_2 + x_5) - \tfrac{1}{6}(2 + 2 + 2)[A B_0] = 0,$$

so that this harmonic point describes the prime $[14.36.25] = 0$, or $[A B_0] = 0$; this again is the polar prime of the node $(A_0 B)$ from which the projection is made.

We thus obtain, corresponding to any node $(\xi_1, \ldots, \xi_6)$ of the primal, the general linear transformation

$$x_i' = x_i - \tfrac{1}{3}\xi_i \varpi,$$

which, when $(x_1, ..., x_6)$ is a point of the primal, leads to another point of the primal. This we shall speak of as a *projection* from the node $(\xi_1, ..., \xi_6)$, and often denote by $p(\xi)$. Herein, ϖ is a linear function of $x_1, ..., x_6$, which, in case $(\xi_1, ..., \xi_6)$ is one of the fifteen nodes (lm), is equal to $3(x_l - x_m)$, or $3[lm]$, and, in case $(\xi_1, ..., \xi_6)$ is one of the thirty nodes (PQ_0), is equal to $[P_0 Q]$. Thus $\varpi = 0$ is the equation of the polar prime of the node $(\xi_1, ..., \xi_6)$, and, when $(x_1, ..., x_6)$, $(x_1', ..., x_6')$ are on the primal, is the locus of the harmonic point of the node in regard to these.

When $(x_1, ..., x_6)$ is a point of a prime passing through the node $(\xi_1, ..., \xi_6)$, the transformed point evidently lies in this prime. When $(x_1, ..., x_6)$ is a node, other than $(\xi_1, ..., \xi_6)$, such that the joining line of these does not lie on the primal, the geometrical interpretation shews that $(x_1', ..., x_6')$ coincides with $(x_1, ..., x_6)$, and it is in accordance with this that the polar prime of $(\xi_1, ..., \xi_6)$ is transformed into itself. When $(x_1, ..., x_6)$ is a node such that its join to $(\xi_1, ..., \xi_6)$ lies on the primal, this joining line being then a κ-line, the algebra does not directly apply. But in fact it may be proved in this case that the point $(x_1', ..., x_6')$ is the third node of the κ-line, so that the polar prime of one node of a κ-line contains the harmonic point of this node in regard to the other two nodes of this κ-line. The forty-five nodes of the primal, other than a chosen node $(\xi_1, ..., \xi_6)$, consist of the twelve nodes lying in the polar prime of $(\xi_1, ..., \xi_6)$, which are all unaltered by the projection, and of the sixteen pairs of nodes lying on the sixteen κ-lines through $(\xi_1, ..., \xi_6)$, of which those of any pair are interchanged; thus a projection gives rise to an even substitution among the nodes of the primal. We add the simple remark that, for all the forty-five nodes, with ϖ either $3(x_l - x_m)$, or $[P_0 Q]$, in the notation above, the substitution of $\xi_1, ..., \xi_6$ respectively for $x_1, ..., x_6$ leads to $\varpi_\xi = 6$. Thus $\varpi_{x'}$ is $\varpi_x - 2\varpi_x$, or $\varpi_{x'} = -\varpi_x$, namely, the linear form whose vanishing gives the polar prime of a node is changed in sign by the projection from that node, but the linear form whose vanishing gives any prime through the node is unaltered by the transformation, because the node lies on this prime.

Evidently, too, in a succession of two transformations which are projections from two angular points of a Jordan pentahedron, the order in which these transformations are carried out is indifferent; and the succession of the five projections from the angular points of the pentahedron leads only to a change of sign of all the coordinates of the projected point, which is thus unaltered.

The projections give forty-five linear transformations of the primal into itself; these are connected by many relations. But it will appear that all linear transformations of the primal into itself are obtainable by combination of such projections.

We have just remarked that the transformations of the primal by projections from two nodes whose join is not a κ-line, are commutable with one another, so that the square of their product is equivalent to identity, and that the composite transformation obtained by projections from all the angular points of a Jordan pentahedron is equivalent to identity. For the relations connecting the forty-five projections, the relations connecting the projections from three nodes which lie on a κ-line are important. If such three nodes be denoted by α, β, γ, and the projections from these be denoted by $p(\alpha)$, $p(\beta)$, $p(\gamma)$, it is found that

$$p(\beta)\,p(\gamma) = p(\gamma)\,p(\alpha) = p(\alpha)\,p(\beta),$$

$$[p(\beta)\,p(\gamma)]^{-1} = p(\gamma)\,p(\beta) = p(\alpha)\,p(\gamma) = p(\beta)\,p(\alpha),$$

$$[p(\beta)\,p(\gamma)]^3 = 1,$$

so that, for instance, $p(\alpha) = p(\beta)\,p(\gamma)\,p(\beta) = p(\gamma)\,p(\beta)\,p(\gamma)$. With the three projections $p(\alpha)$, $p(\beta)$, $p(\gamma)$, we thus obtain a group of six transformations, isomorphic with the substitutions of three letters. These statements are immediately obvious for three nodes (jk), (ki), (ij), but can be verified by the general formula for both the other types of κ-line enumerated under (h), (c) in § (5).

(8)   **The forty Steiner threefold spaces, or primes, belonging to the primal.** There are forty solids, or primes, whose intersections with the Burkhardt primal consist each of a quartic

surface which breaks up into four planes. These planes are then
Jacobian planes. The equations of these primes are of one of the
two forms $x_l + x_m + x_n = 0$, $x_l - \epsilon x_m = 0$. The former equation, if
l, m, n, p, q, r be the numbers $1, 2, ..., 6$ in some order, is equivalent
with $x_p + x_q + x_r = 0$, and there are ten primes with such equations;
of primes with equations of the form $x_l - \epsilon x_m = 0$ there are 2.15,
or 30, the two numbers l, m giving rise to two $x_l - \epsilon x_m = 0$ and
$x_m - \epsilon x_l = 0$, of which the latter is $x_l - \epsilon^2 x_m = 0$. Notwithstanding
the difference in the forms of the equations, these forty primes are
of equal standing; to see this we may employ the self-transforma-
tion of the primal $(x') = \chi(x)$ already used in § (3); this leads to

$$x_1' - \epsilon x_4' = x_1 + \epsilon x_2 + \epsilon x_3 + \epsilon x_4 + x_5 + x_6 = (\epsilon - 1)(x_4 + x_2 + x_3),$$

and

$$x_4' + x_2' + x_3' = 2\epsilon x_1 - \epsilon^2 x_2 - \epsilon^2 x_3 - x_4 - \epsilon^2 x_5 - \epsilon^2 x_6 = (\epsilon - 1)(x_1 - \epsilon^2 x_4),$$

which are sufficient to shew that there is no difference in character
between primes of the form $x_l + x_m + x_n = 0$, and primes of the
form $x_l - \epsilon x_m = 0$. This transformation also gives, for instance,

$$x_1' + x_2' + x_4' = (\epsilon^2 - 1)(x_4 - \epsilon x_3).$$

The four planes in which any one of the forty solids meets the
primal form a tetrahedron in this solid. For reasons referred to
below, we call any of these solids a *Steiner solid*, and the tetrahedron therein a *Steiner tetrahedron*.

As an example of the statement made, which is also effectively
a proof of it, consider the solid $x_1 + x_2 + x_3 = 0$. This prime
evidently contains the four points F, R, L, T of respective
coordinates

$$F(0, 0, 0, 1, \epsilon, \epsilon^2); \quad R(1, \epsilon, \epsilon^2, 0, 0, 0);$$

$$L(1, \epsilon^2, \epsilon, 0, 0, 0); \quad T(0, 0, 0, 1, \epsilon^2, \epsilon),$$

which form a tetrahedron, the faces RTL, TLF, TFR, FRL
being the respective planes

$$x_4/1 = x_5/\epsilon^2 = x_6/\epsilon; \quad x_1/1 = x_2/\epsilon^2 = x_3/\epsilon;$$

$$x_1/1 = x_2/\epsilon = x_3/\epsilon^2; \quad x_4/1 = x_5/\epsilon = x_6/\epsilon^2,$$

which are Jacobian planes, lying on the primal; these meet in edges, TL, FR; TR, LF; LR, TF containing, respectively, the nodes

$$(BC_0), (CA_0), (AB_0); \quad (B_0C), (C_0A), (A_0B); \quad (EF_0), (FD_0), (DE_0);$$

$$(E_0F), (F_0D), (D_0E); \quad (23), (31), (12); \quad (56), (64), (45),$$

as we immediately find from the fundamental scheme of notations in § (1). The opposite edges of the tetrahedron are thus polar κ-lines, by what we have seen, and any pair of opposite edges determine the solid. Indeed, any one of the edges, being a κ-line, determines the opposite polar κ-line, and then the pairs of Jacobian planes passing through these two edges determine the tetrahedron, and the other two pairs of opposite edges.

As another example, consider the solid whose equation is $x_5 - \epsilon x_3 = 0$. It is clear that this contains the four points F, R, L, T, of respective coordinates

$$F(1, 1, \epsilon, -2, \epsilon^2, 1); \quad R(-2, 1, \epsilon, 1, \epsilon^2, 1);$$

$$L(1, 1, \epsilon, 1, \epsilon^2, -2); \quad T(1, -2, \epsilon, 1, \epsilon^2, 1),$$

forming a tetrahedron with the respectively opposite plane faces RLT, LTF, FRT, FRL given by

$$x_3/1 = x_4/\epsilon^2 = x_5/\epsilon; \quad x_1/1 = x_3/\epsilon = x_5/\epsilon^2;$$

$$x_3/1 = x_5/\epsilon = x_6/\epsilon^2; \quad x_2/1 = x_3/\epsilon = x_5/\epsilon^2,$$

which are Jacobian planes, lying on the primal. The three pairs of opposite edges TL, FR; TR, LF; LR, TF are then pairs of polar κ-lines containing respectively the nodes

$$(AB_0), (CD_0), (26); \quad (AD_0), (CB_0), (14); \quad (EC_0), (FA_0), (12);$$

$$(EA_0), (FC_0), (46); \quad (BF_0), (DE_0), (16); \quad (BE_0), (DF_0), (24),$$

as we easily verify by the scheme given in § (1).

The 240 existing κ-lines thus appear as distributed in sets of six in the forty Steiner tetrahedra, no two of which, as we have seen, have a common edge. Also it is easy to see, in each case, that the pair of Hessian points of the three nodes lying on an edge of a

Steiner tetrahedron are the two angular points of the tetrahedron lying on that edge; in any Steiner space, if $X = 0$, $Y = 0$, $Z = 0$, $T = 0$ be the equations of the faces of the tetrahedron, the eighteen nodes lying on the edges of the tetrahedron lie in fact upon a cubic surface whose equation may be taken to be

$$X^3 + Y^3 + Z^3 + T^3 = 0.$$

The twenty-seven lines of this surface are the lines joining the three nodes in any edge each to all the nodes of the opposite edge. Thus the eighteen nodes lie in sixes upon twenty-seven planes; these planes do not lie on the Burkhardt primal; each meets the primal in a quartic curve which breaks up into four κ-lines (see §(9) following).

The Steiner solids arise differently below (§(13)), where we express the equation of the Burkhardt primal in terms of five primes, of which one is a Steiner solid, taken with four associated solids, in forty different ways. From the new equation it is instantaneously obvious that every Steiner solid is a tangent prime of the primal, and of a singular character, in that *the solid is the tangent prime at every ordinary point of every one of the six edges of the Steiner tetrahedron contained therein.* This may also be verified with the equation used here.

(9) **The plane common to two Steiner solids.** In every one of the four faces of a Steiner tetrahedron, in addition to the triangle formed by the edges in that plane, there are (§(4)), three other triangles, of which the sides are κ-lines, so that all the nine nodes of the plane lie on the sides of any one of the four triangles. Through each side of such a triangle, besides the plane of the face considered, there passes another Jacobian plane. It can be proved that the three new planes so arising from any one of the three additional triangles in this plane face, form with this plane, the faces of another Steiner tetrahedron. Thus there are twelve Steiner solids each having a plane, lying on the primal, in common with a given Steiner solid, three of these for each of the planes of the given Steiner tetrahedron. For instance, the

Steiner solid $x_2 - \epsilon x_3 = 0$, evidently has, common with the solid $x_1 + x_2 + x_3 = 0$, the plane $x_1/1 = x_2/\epsilon^2 = x_3/\epsilon$. And it can be shewn that the transformation, which we denote by $(x') = B(x)$, given by

$$x_1' = x_1 + \epsilon x_3 + \epsilon x_6, \qquad -x_3' = x_2 + \epsilon^2 x_4 + \epsilon^2 x_5,$$

$$x_5' = \epsilon x_1 + x_3 + \epsilon x_6, \qquad -x_4' = \epsilon^2 x_2 + x_4 + \epsilon^2 x_5,$$

$$x_2' = \epsilon x_1 + \epsilon x_3 + x_6, \qquad -x_6' = \epsilon^2 x_2 + \epsilon^2 x_4 + x_5$$

(which evidently leaves the primal unaltered, because it is obtainable from the transformation $(x') = \chi(x)$, of § (3), by suitable substitutions among $x_1, \ldots, x_6$, and suitable substitutions among $(x_1', \ldots, x_6')$) leads to

$$x_1' + x_2' + x_3' = (\epsilon^2 - 1)(x_2 - \epsilon x_3), \quad x_2' - \epsilon x_3' = (\epsilon - 1)(x_1 + x_2 + x_3).$$

Similarly, the Steiner solid $x_1 + x_2 + x_3 = 0$, which is entirely general, has a Jacobian plane in common with each of the twelve solids $x_l - \epsilon x_m = 0$, $x_l - \epsilon^2 x_m = 0$ (l, m being any two of the numbers 1, 2, 3, or any two of the numbers 4, 5, 6).

There remain then $40 - 12 - 1$, or 27, Steiner solids which have not a Jacobian plane in common with a given Steiner solid, say $x_1 + x_2 + x_3 = 0$. These consist in fact of the nine solids

$$x_p + x_l + x_m = 0,$$

where l, m are two of the numbers 1, 2, 3 and p is one of the numbers 4, 5, 6, together with the eighteen solids, $x_l - \epsilon x_p = 0$, $x_l - \epsilon^2 x_p = 0$, where l is one of 1, 2, 3 and p is one of 4, 5, 6. *Any two such Steiner solids, intersecting in a plane which is not a Jacobian plane, with their associated Steiner tetrahedra, are in fact in perspective with one another from one of the forty-five nodes, and their common plane lies on the Jordan prime which is the polar of this node.* For example, we may consider first the case of the two Steiner solids $x_1 + x_2 + x_3 = 0$, $x_4 + x_2 + x_3 = 0$, and then the case of the two solids $x_1 + x_2 + x_3 = 0$, $x_1 - \epsilon x_4 = 0$. Evidently, the two first solids meet on the Jordan prime $[14] = 0$, and are projections of one another from the node (14). The prime $[14] = 0$

contains the twelve angular points of the pentahedra $\{AB\}$, $\{CD\}$, $\{EF\}$, which have (14) for common angular point, namely,

$$(AB_0), (A_0B), (36), (25); \quad (CD_0), (C_0D), (23), (56);$$

$$(EF_0), (E_0F), (26), (35).$$

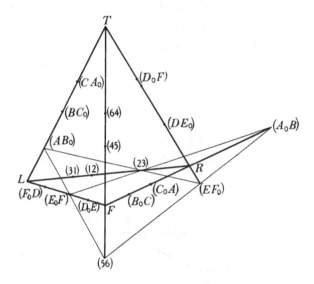

Of these twelve nodes there are six lying in both $x_1 + x_2 + x_3 = 0$ and $x_4 + x_2 + x_3 = 0$, these being the intersections of four κ-lines lying in the common plane of the two solids, namely,

$$(A_0B), (E_0F), (23); \quad (A_0B), (EF_0), (56);$$

$$(AB_0), (E_0F), (56); \quad (AB_0), (EF_0), (23),$$

and this common plane does not lie on the primal.

Considering next $x_1 + x_2 + x_3 = 0$ and $x_1 - \epsilon x_4 = 0$, we verify that

$$(\epsilon^2 - \epsilon)[x_1 + x_2 + x_3 + \epsilon(x_1 - \epsilon x_4)]$$
$$= x_1 + x_4 + \epsilon(x_5 + x_6) + \epsilon^2(x_2 + x_3), = [C_0D];$$

hence these two Steiner solids meet in a plane lying on the Jordan prime $[C_0D] = 0$; also $x_1 + x_2 + x_3 - \epsilon(x_1 - \epsilon x_4)$ vanishes at the pole (CD_0), or $(14.23.56)$. Thus $[C_0 D] = 0$ is the harmonic prime of (CD_0) in regard to the two Steiner solids. Further, the six nodes

$$(CA_0),\ (CB_0),\ (D_0 E),\ (D_0 F),\ (23),\ (56),$$

which lie in $[C_0 D] = 0$, are easily seen to lie both in $x_1 + x_2 + x_3 = 0$ and in $x_1 - \epsilon x_4 = 0$, and to be the intersections of the four κ-lines

$$(CB_0),\ (D_0 E),\ (23);\quad (CB_0),\ (D_0 F),\ (56);$$
$$(CA_0),\ (D_0 E),\ (56);\quad (CA_0),\ (D_0 F),\ (23),$$

whose plane, common to the two solids, does not lie on the primal.

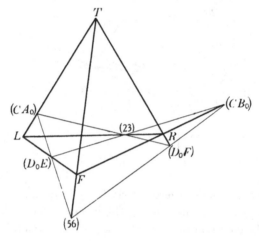

Conversely, take the Steiner tetrahedron in the solid

$$x_1 + x_2 + x_3 = 0,$$

and one of the nodes, say $(A_0 B)$, in the edge containing $(A_0 B)$, $(B_0 C)$, $(C_0 A)$. Through this node can be drawn a κ-line, besides this edge, in each of the two faces of the tetrahedron which meet in this edge; say these are $(A_0 B)$, (23), $(E_0 F)$ and $(A_0 B)$, (EF_0), (56). In the plane of these two lines the κ-lines joining (EF_0), (23) and (56), $(E_0 F)$ must meet; and they meet in the node (AB_0) lying on the edge of the tetrahedron opposite to that on which $(A_0 B)$ is taken. The diagonals of the quadrilateral so formed by the four κ-lines, namely the lines (23), (56); (AB_0), (A_0B); (EF_0), (E_0F), do not lie on the primal, but each is an edge of a Jordan pentahedron, respectively $\{CD\}$, $\{AB\}$, $\{EF\}$; and these pentahedra have a common angular point, namely (14), whose polar prime contains the four κ-lines; the solid $x_1 + x_2 + x_3 = 0$ is projected from (14) into $x_4 + x_2 + x_3 = 0$, which equally contains these four κ-lines. As the three nodes (A_0B), (23), (EF_0), which determine the construction are, each, one of three in the edge in which it lies, the construction can be made in twenty-seven ways, and leads to the twenty-seven Steiner solids which meet

$$x_1 + x_2 + x_3 = 0$$

in a plane not lying on the primal; these twenty-seven solids are the projections of $x_1 + x_2 + x_3 = 0$ from the twenty-seven nodes other than the eighteen nodes which lie on the edges of the tetrahedron belonging to $x_1 + x_2 + x_3 = 0$.

Such a plane of intersection of two Steiner solids which have no Jacobian plane in common, may, for distinctness, be called a *cross-plane*. By what we have seen, the number of such planes is $\frac{1}{2} 40.27$, or 540. Each lies in a particular Jordan prime, so that the number in a particular Jordan prime is $540 \div 45$, or 12. The three diagonals of the quadrilateral of κ-lines in such a plane are respective edges of the three tetrahedra formed by the nodes in the Jordan prime, and, as each tetrahedron has six edges, there are two of the cross-planes through every one of the eighteen edges of the three tetrahedra; in fact, for three tetrahedra in desmic relation, in a threefold space, there are twelve planes each containing an edge of all three tetrahedra, two of these planes through every edge.

We refrain from further elaboration of the theory of the cross-planes, making only the following remarks: (1) a Steiner solid contains forty-two κ-lines, of which seven pass through each of the eighteen nodes; (2) through every κ-line there pass seven Steiner solids, of which only one has this κ-line as an edge of the associated Steiner tetrahedron; this is in agreement with

$7.240 = 40.42$; (3) there are sixteen Steiner solids containing any node, the remaining twenty-four Steiner solids being in perspective in pairs from that node, but through two nodes whose join is not a κ-line there pass four Steiner solids; (4) the four κ-lines of a cross-plane have for polars four κ-lines which all contain the node which is the pole of the Jordan prime containing the cross-plane.

(10) **The enumeration of the twenty-seven Jordan pentahedra, and of the forty-five nodes, from the nodes in pairs of polar x-lines.** The nodes in one edge of a Steiner tetrahedron can be joined to the nodes in the opposite edge by nine lines. These are not κ-lines, but each is an edge-line of a definite Jordan pentahedron. We illustrate this by considering the three pairs of opposite edges of the Steiner tetrahedron considered in the first example of § (8). Each of these pairs of edges gives rise to edge-lines of nine pentahedra

	(BC_0)	(CA_0)	(AB_0)		(EF_0)	(FD_0)	(DE_0)		(56)	(64)	(45)
(B_0C)	$\{BC\}$	$\{C\}$	$\{B_0\}$	(E_0F)	$\{EF\}$	$\{F\}$	$\{E_0\}$	(23)	$\{CD\}$	$\{AF\}$	$\{BE\}$
(C_0A)	$\{C_0\}$	$\{CA\}$	$\{A\}$	(F_0D)	$\{F_0\}$	$\{FD\}$	$\{D\}$	(31)	$\{BF\}$	$\{CE\}$	$\{AD\}$
(A_0B)	$\{B\}$	$\{A_0\}$	$\{AB\}$	(D_0E)	$\{E\}$	$\{D_0\}$	$\{DE\}$	(12)	$\{AE\}$	$\{BD\}$	$\{CF\}$

and the schemes put down shew what these pentahedra are. For instance, in the first scheme (BC_0), (B_0C) are angular points of the pentahedron $\{BC\}$, while (CA_0), (B_0C) are angular points of the pentahedron $\{C\}$.

We see that all the twenty-seven pentahedra arise from the aggregate of the three schemes. Moreover, each of these pentahedra has three nodes besides the two by which it is enumerated; thus, if we take one pair of opposite edges, say those in the first scheme, we obtain 9.3 or twenty-seven nodes besides those in the two edges. If we consider the second pair of opposite edges, and the second scheme, we likewise obtain twenty-seven new nodes, but these are the same as those arising from the first pair of opposite edges. The third pair of opposite edges likewise give the same twenty-seven nodes. And these twenty-seven, in fact, are those of the forty-five nodes other than those on the six edges of the tetrahedron. Thus all the forty-five nodes arise from the chosen Steiner tetrahedron. A similar remark, and schemes, arise for the other Steiner tetrahedron considered in § (8):

	(AB_0)	(CD_0)	(26)		(BE_0)	(DF_0)	(24)		(EA_0)	(FC_0)	(46)
(AD_0)	$\{A\}$	$\{D_0\}$	$\{AD\}$	(BF_0)	$\{B\}$	$\{F_0\}$	$\{BF\}$ (EC_0)		$\{E\}$	$\{C_0\}$	$\{CE\}$
(CB_0)	$\{B_0\}$	$\{C\}$	$\{BC\}$	(DE_0)	$\{E_0\}$	$\{D\}$	$\{DE\}$ (FA_0)		$\{A_0\}$	$\{F\}$	$\{AF\}$
(14)	$\{AB\}$	$\{CD\}$	$\{EF\}$	(16)	$\{BE\}$	$\{DF\}$	$\{AC\}$ (12)		$\{AE\}$	$\{CF\}$	$\{BD\}$

(11) **The reason for calling the Steiner tetrahedra by this name.** This seems the appropriate place to explain why we have spoken of κ-lines, and of Steiner tetrahedra.

First, remark a theorem of elementary geometry in ordinary threefold space. Suppose we have three triangles, of angular points $A_1 B_1 C_1$, $A_2 B_2 C_2$, $A_3 B_3 C_3$, and respectively opposite sides $a_1 b_1 c_1$, $a_2 b_2 c_2$, $a_3 b_3 c_3$, which are such that the points $A_1 A_2 A_3$ are in line, as also the points $B_1 B_2 B_3$ and the points $C_1 C_2 C_3$, these three lines meeting in a point O, so that the three triangles are in perspective from O. Then, the three sides $a_1 a_2 a_3$, all lying in the plane $OB_1 C_1 B_2 C_2 B_3 C_3$ form a triangle, as do the sides $b_1 b_2 b_3$ in the plane $OC_1 A_1 C_2 A_2 C_3 A_3$, and the sides $c_1 c_2 c_3$ in the plane $OA_1 B_1 A_2 B_2 A_3 B_3$, and the planes of these triangles meet in O. The planes of the triangles $A_1 B_1 C_1$, $A_2 B_2 C_2$, $A_3 B_3 C_3$ also meet in a point, say Q. The theorem referred to is that, *the triangles $a_1 a_2 a_3$, $b_1 b_2 b_3$, $c_1 c_2 c_3$ are in perspective, from the point Q.* To see, for example, that the two triangles $b_1 b_2 b_3$, $c_1 c_2 c_3$ are in perspective, it is sufficient to see that the pairs of sides, from these triangles, b_1 and c_1, b_2 and c_2, b_3 and c_3, meet, and thus meet in three points lying on the line of intersection of the planes of these two triangles; in fact, these points of meeting are respectively A_1, A_2, A_3. It follows that the three lines joining corresponding angular points of these two triangles meet in a point; these lines join respectively the points $(b_2 b_3)$, $(c_2 c_3)$; the points $(b_3 b_1)$, $(c_3 c_1)$; and the points $(b_1 b_2)$, $(c_1 c_2)$. But the plane containing the latter

3-2

two of these intersecting lines contains the lines b_1 and c_1, and is the plane of the triangle $A_1 B_1 C_1$. The centre of perspective of the triangles $b_1 b_2 b_3$, $c_1 c_2 c_3$ thus lies in the plane $A_1 B_1 C_1$. By parity of reasoning it is thus common to the three planes $A_1 B_1 C_1$, $A_2 B_2 C_2$, $A_3 B_3 C_3$, and is the point Q. This point is similarly a centre of perspective for each pair of the three triangles $a_1 a_2 a_3$, $b_1 b_2 b_3$, $c_1 c_2 c_3$. This is what we desired to prove.

Now consider a general cubic surface in ordinary threefold space. Suppose that, on this surface, there are two sets of three lines, $a_1 b_1 c_1$ and $a_2 b_2 c_2$, forming two triangles with no side in common. The sides of these two triangles all meet the line of intersection of the planes of the triangles, and all points of every one of these sides are on the cubic surface. Thus, either the line of intersection of the planes of the two triangles lies on the cubic surface, which is impossible, since then the plane of either triangle would meet the cubic surface in a (composite) curve of the fourth order, or else the six points in which this line is met by the sides of the two triangles are coincident, in sets, in the three points in which this line meets the cubic surface. We suppose the symmetrical general case to arise, namely that these six points coincide in pairs in these three points. The two triangles $a_1 b_1 c_1$, $a_2 b_2 c_2$ are thus in perspective, with the line of intersection of their planes as axis of perspective; we suppose the notation chosen so that $a_1 a_2$ meet on this line, as also $b_1 b_2$ and $c_1 c_2$. It follows then, if $A_1 B_1 C_1$, $A_2 B_2 C_2$ be the angular points of these two triangles, that the lines $A_1 A_2$, $B_1 B_2$, $C_1 C_2$ meet in a point, say O. Now each of these lines meets the cubic surface in a further point, say, respectively, A_3, B_3, C_3. And the plane $O B_1 B_2 C_1 C_2$, meeting the cubic surface in the two lines a_1, a_2, will meet the surface in a third line, namely the line, say a_3, which joins B_3 and C_3. The triangle $A_3 B_3 C_3$ has thus, for sides, three other lines of the cubic surface, say a_3, b_3, c_3. We thus arrive at the figure just considered, of three triangles $A_1 B_1 C_1$, $A_2 B_2 C_2$, $A_3 B_3 C_3$, in perspective from O. And the nine lines $a_1 a_2 a_3$, $b_1 b_2 b_3$, $c_1 c_2 c_3$ form three other triangles, in perspective from a point Q. These nine lines on the cubic surface are then the intersections of three planes meeting in O with three planes meeting in Q.

These results are in Steiner's paper on Cubic Surfaces (*Ges. Werke*, II (1856), p. 655). He names a line which contains three intersections, each of two lines of the surface, a k-line, and obtains the result that, through any intersection of two lines of the surface, there pass sixteen k-lines.

The Burkhardt primal is in fact in correspondence with a cubic surface, in the sense that the plane of any three coplanar lines of the cubic surface corresponds to a node of the primal. And two sets of three planes meeting in nine lines of the surface, such as those considered above, correspond to the two sets of three nodes of the primal lying on two opposite edges of a Steiner tetrahedron, the nine lines corresponding to the pentahedra of which the joins of the two sets of three nodes are edge-lines, as in § (10). These two opposite edges, which we have called polar κ-lines, determine the tetrahedron, and so determine the other two pairs of opposite edges of the tetrahedron. Likewise, for the cubic surface, given nine lines forming two sets of three triangles, each in perspective, as above, it is possible to determine the remaining eighteen lines of the surface, and to see that these consist of two other such sets of nine lines. Also, for the cubic surface, there are forty ways of arranging the twenty-seven lines in three sets of nine, each set forming two batches of three triangles in perspective, as above. This corresponds to the existence of forty Steiner tetrahedra on the primal.

(12) **The enumeration of the twenty-seven pentahedra from nine nodes of the Burkhardt primal.** Resuming the matter discussed in § (10), it appears that, if we take a Steiner tetrahedron, and from each of the three pairs of opposite edges we choose one edge, thus selecting nine nodes of the primal, and then consider the three Jordan pentahedra of which each of these nine nodes is an angular point, then we shall obtain all the twenty-seven pentahedra.

In particular, taking three edges of the Steiner tetrahedron which lie in a plane, it appears that all the twenty-seven pentahedra may be enumerated as those which have angular points, in threes, at the nine nodes of any Jacobian plane. For instance, taking the former Steiner tetrahedron considered in §(10), we can arrange the twenty-seven pentahedra in nine columns:

(BC_0)	(CA_0)	(AB_0)	(EF_0)	(FD_0)	(DE_0)	(23)	(31)	(12)
$\{BC\}$	$\{C\}$	$\{B_0\}$	$\{EF\}$	$\{F\}$	$\{E_0\}$	$\{CD\}$	$\{BF\}$	$\{AE\}$
$\{C_0\}$	$\{CA\}$	$\{A\}$	$\{F_0\}$	$\{FD\}$	$\{D\}$	$\{AF\}$	$\{CE\}$	$\{BD\}$
$\{B\}$	$\{A_0\}$	$\{AB\}$	$\{E\}$	$\{D_0\}$	$\{DE\}$	$\{BE\}$	$\{AD\}$	$\{CF\}$

of which those in a column have the common angular point written above, and these angular points are the nodes of a Jacobian plane $(x_4/1 = x_5/\epsilon^2 = x_6/\epsilon)$. There are forty such possibilities. On the cubic surface, as in §(11), any line is met by five pairs of intersecting lines, and so belongs to five sets of three coplanar lines; the lines of the cubic surface correspond in this sense to the twenty-seven Jordan pentahedra of the Burkhardt primal. The scheme thus determines one of forty ways in which the twenty-seven lines of the cubic surface can be obtained by intersection of the surface with nine planes.

Or, we may take a scheme in which the nine nodes employed are those in three concurrent edges of a Steiner tetrahedron, of which there are then 160 cases. For example, we may take the nine nodes to be

$$(BC_0), (CA_0), (AB_0), (EF_0), (FD_0), (DE_0), (56), (64), (45);$$

we thus obtain the same scheme as that written above except that the three last columns of pentahedra then arising are those written in the *rows* under (23), (31), (12) in the above scheme.

(13) **The equation of the Burkhardt primal in terms of a Steiner solid and four associated primes.** We have stated (§(8)) that there are forty primes, the Steiner solids, which meet the primal each in four planes. We shew now how to associate with every Steiner solid, four other primes, not them-

selves Steiner solids, forming with this Steiner solid a simplex, and obtain the equation of the primal referred to such a simplex. The equation obtained shews at once that the Steiner solid meets the four associated primes in planes which lie on the primal.

For this purpose we introduce five primes,

$$y_0 = 0, y_1 = 0, \ldots, y_4 = 0,$$

which are given, in terms of $x_1, x_2, \ldots, x_6$, by the equations

$$3y_0 = x_1 + x_2 + x_3, \qquad -3y_0 = x_4 + x_5 + x_6,$$
$$3y_1 = x_1 + \epsilon^2 x_2 + \epsilon x_3, \qquad -3y_2 = x_4 + \epsilon^2 x_5 + \epsilon x_6,$$
$$3y_4 = x_1 + \epsilon x_2 + \epsilon^2 x_3, \qquad -3y_3 = x_4 + \epsilon x_5 + \epsilon^2 x_6,$$

of which the reverse equations are

$$x_1 = y_0 + y_1 + y_4, \qquad -x_4 = y_0 + y_2 + y_3,$$
$$x_2 = y_0 + \epsilon y_1 + \epsilon^2 y_4, \qquad -x_5 = y_0 + \epsilon y_2 + \epsilon^2 y_3,$$
$$x_3 = y_0 + \epsilon^2 y_1 + \epsilon y_4, \qquad -x_6 = y_0 + \epsilon^2 y_2 + \epsilon y_3.$$

These equations, *which are constantly employed*, shew that $y_0 = 0, y_1 = 0, \ldots, y_4 = 0$ form a simplex, of which $y_0 = 0$ is a Steiner solid. In terms of these the equation of the Burkhardt primal, as shewn below, takes the form

$$y_0^4 - y_0(y_1^3 + y_2^3 + y_3^3 + y_4^3) + 3y_1 y_2 y_3 y_4 = 0; \qquad \text{(I)};$$

this shews that $y_0 = 0$ meets the primal in the four planes $y_0 = y_1 = 0$; $y_0 = y_2 = 0$; $y_0 = y_3 = 0$; $y_0 = y_4 = 0$. It is easy to see that these four planes are, respectively,

$$x_1/1 = x_2/\epsilon^2 = x_3/\epsilon; \quad x_4/1 = x_5/\epsilon^2 = x_6/\epsilon;$$
$$x_4/1 = x_5/\epsilon = x_6/\epsilon^2; \quad x_1/1 = x_2/\epsilon = x_3/\epsilon^2.$$

To verify the new form of the equation of the primal, we remark that

$$x_1 x_2 x_3 = y_0^3 + y_1^3 + y_4^3 - 3y_0 y_1 y_4,$$
$$-x_4 x_5 x_6 = y_0^3 + y_2^3 + y_3^3 - 3y_0 y_2 y_3,$$
$$x_2 x_3 + x_3 x_1 + x_1 x_2 = 3(y_0^2 - y_1 y_4),$$
$$x_5 x_6 + x_6 x_4 + x_4 x_5 = 3(y_0^2 - y_2 y_3);$$

the equation of the primal, in $x_1, \ldots, x_6$, is

$$x_1 x_2 x_3 (x_4 + x_5 + x_6) + x_4 x_5 x_6 (x_1 + x_2 + x_3)$$
$$+ (x_2 x_3 + x_3 x_1 + x_1 x_2)(x_5 x_6 + x_6 x_4 + x_4 x_5) = 0,$$

and substitution leads, save for the factor 3, to

$$-y_0(y_0^3 + y_1^3 + y_4^3 - 3y_0 y_1 y_4) - y_0(y_0^3 + y_2^3 + y_3^3 - 3y_0 y_2 y_3)$$
$$+ 3(y_0^2 - y_1 y_4)(y_0^2 - y_2 y_3) = 0,$$

which is evidently equivalent to the equation (I) given above.

Further, as this equation holds solely in virtue of the equation connecting $x_1, \ldots, x_6$, between

$$x_1 + x_2 + x_3, \quad x_1 + \epsilon^2 x_2 + \epsilon x_3, \quad -(x_4 + \epsilon^2 x_5 + \epsilon x_6),$$
$$-(x_4 + \epsilon x_5 + \epsilon^2 x_6), \quad x_1 + \epsilon x_2 + \epsilon^2 x_3,$$

it follows, by the symmetry of the original equation, that a precisely identical equation holds for the five linear forms in $x_1, \ldots, x_6$, obtained from the previous five forms by interchange of x_1 and x_4, namely between $\eta_0, \eta_1, \ldots, \eta_4$, given by

$$\eta_0 = x_4 + x_2 + x_3, \qquad \eta_2 = -(x_1 + \epsilon^2 x_5 + \epsilon x_6),$$
$$\eta_1 = x_4 + \epsilon^2 x_2 + \epsilon x_3, \qquad \eta_3 = -(x_1 + \epsilon x_5 + \epsilon^2 x_6),$$
$$\eta_4 = x_4 + \epsilon x_2 + \epsilon^2 x_3.$$

By means of the equations connecting $x_1, \ldots, x_6$ and $y_0, y_1, \ldots, y_4$, these five new forms are found to be

$$\eta_0 = y_0 - y_1 - y_2 - y_3 - y_4, \quad \eta_1 = -2y_0 + 2y_1 - y_2 - y_3 - y_4, \text{ etc.},$$

or, in general form,

$$\eta_i = 3(y_i - \tfrac{1}{3}u), \quad i = 0, 1, 2, 3, 4,$$

where $u = 2y_0 + y_1 + y_2 + y_3 + y_4$. These are, in fact, as we may verify directly from equation (I), the equations for the projection of the primal from the node (14), for which $y_0 = y_1 = y_2 = y_3 = y_4$ (cf. § (7)).

Further still, the equation (I) shews, on trial, that an equation, of precisely the same form,

$$y_0'^4 - y_0'(y_1'^3 + y_2'^3 + y_3'^3 + y_4'^3) + 3y_1' y_2' y_3' y_4' = 0$$

holds for

$$y_0' = y_0 - y_1, \qquad y_1' = -2y_0 - y_1,$$

$$y_2' = y_2 + y_3 + y_4, \quad y_3' = y_2 + \epsilon y_3 + \epsilon^2 y_4, \quad y_4' = y_2 + \epsilon^2 y_3 + \epsilon y_4.$$

This fact is clear also, if we recall (§ (3)) that the original equation of the primal, in $x_1, ..., x_6$, is unaltered by the linear transformation there denoted by $(x') = \chi(x)$; for, this being so, it follows from the symmetry of the original equation, that this equation is also unaltered by the linear transformation (already used in § (9))

$$x_1' = x_1 + \epsilon x_3 + \epsilon x_6, \quad -x_3' = x_2 + \epsilon^2 x_4 + \epsilon^2 x_5,$$

$$x_5' = \epsilon x_1 + x_3 + \epsilon x_6, \quad -x_4' = \epsilon^2 x_2 + x_4 + \epsilon^2 x_5,$$

$$x_2' = \epsilon x_1 + \epsilon x_3 + x_6, \quad -x_6' = \epsilon^2 x_2 + \epsilon^2 x_4 + x_5,$$

and, replacing $x_1, ..., x_6$ by $y_0, y_1, ..., y_4$, by the fundamental formulae here introduced, and replacing $x_1', ..., x_6'$ by $y_0', y_1', ..., y_4'$, by the same formulae, this transformation is that put down, connecting $y_0, y_1, ..., y_4$ and $y_0', y_1', ..., y_4'$.

The three results thus obtained shew, however, that there are forty sets of values for $y_0, y_1, ..., y_4$, all linear functions of the original $y_0, y_1, ..., y_4$, which are connected by precisely the same equation (I). The forms $y_0 = x_1 + x_2 + x_3$, $\eta_0 = x_4 + x_2 + x_3$, $y_0' = y_0 - y_1$ (which, save for a constant factor, is $x_2 - \epsilon x_3$), are such that $y_0 = 0$, $\eta_0 = 0$, $y_0' = 0$ are all Steiner solids, and we have shewn that there are forty such, all similar. The corresponding results for $y_0, y_1, ..., y_4$ are derivable by the remark that the equation (I) is unaltered by interchange of y_1, y_2, y_3, y_4 among themselves, y_0 remaining unchanged, and is also unaltered by replacing y_1, y_2, y_3, y_4 respectively by $\epsilon^\alpha y_1$, $\epsilon^\beta y_2$, $\epsilon^\gamma y_3$, $\epsilon^\delta y_4$, where each of $\alpha, \beta, \gamma, \delta$ is 0, or 1, or 2, subject to $\alpha + \beta + \gamma + \delta \equiv 0$ (mod. 3). Thus, recalling the form of η_0, in terms of $y_0, y_1, ..., y_4$, there is an equation of the form (I) in which y_0 is replaced by

$$y_0 - \epsilon^\alpha y_1 - \epsilon^\beta y_2 - \epsilon^\gamma y_3 - \epsilon^\delta y_4,$$

with appropriate corresponding replacements for y_1, y_2, y_3, y_4, and this gives twenty-seven forms for η_0. And, recalling the form of y_0', in terms of $y_0, y_1, \ldots, y_4$, there is an equation of the form (I) in which y_0 is replaced by $y_0 - \epsilon^\lambda y_i$, for $\lambda = 0, 1, 2$ and $i = 1, 2, 3, 4$, with appropriate corresponding replacements for $y_1, \ldots, y_4$, and this gives twelve forms for y_0'. The total number of equations (I) thus obtainable is $1 + 27 + 12$ or forty, in each of which the form of y_0 gives a Steiner solid.

The left side of equation (I) is the simplest of the invariants which Burkhardt obtains for the transformation of the theta functions he considers, arising by transformation of the periods of the functions.

Some illustrative remarks may be added in regard to the transformations here considered:

(a) The inverse of the transformation by which we pass from $y_0, y_1, \ldots, y_4$ to $y_0', y_1', \ldots, y_4'$, may easily be computed. Save for a common factor 3, of all the coordinates, which we neglect, this inverse transformation has the same form as the direct transformation, save that, in the transformed values of y_3 and y_4, in place of ϵ, ϵ^2, there occur ϵ^2, ϵ. If then the transformation be denoted (as in §(9)) by $(y_0', \ldots, y_4') = B(y_0, \ldots, y_4)$, the equations $(y_0''', \ldots, y_4''') = B^{-1}(y_0'', \ldots, y_4'')$ are

$$y_0''' = y_0'' - y_1'', \qquad y_1''' = -2y_0'' - y_1'',$$

$$y_2''' = y_2'' + y_3'' + y_4'', \quad y_3''' = y_2'' + \epsilon^2 y_3'' + \epsilon y_4'', \quad y_4''' = y_2'' + \epsilon y_3'' + \epsilon^2 y_4''.$$

(b) By the general formula for projection from a node (§(7)), we can put down the form for projection from the node $(A_0 B)$, from variables $x_1, \ldots, x_6$ to variables $x_1', \ldots, x_6'$. If we replace $(x_1, \ldots, x_6)$ by $(y_0, y_1, \ldots, y_4)$, by the definitions above (§(13)), and, *by the same equations*, replace $(x_1', \ldots, x_6')$ by $(y_0', y_1', \ldots, y_4')$, we find that $y_0', \ldots, y_4'$ are respectively the same as $y_0, \ldots, y_4$ save that $y_1' = y_2$, $y_2' = y_1$. This transformation we denote by D, so that $(y_0'', \ldots, y_4'') = D(y_0', \ldots, y_4')$ is given by

$$y_0'' = y_0', \quad y_1'' = y_2', \quad y_2'' = y_1', \quad y_3'' = y_3', \quad y_4'' = y_4'.$$

(c) If we now express $y_0''', ..., y_4'''$ in terms of $y_0, ..., y_4$ by means of the equations

$$(y_0''', ..., y_4''') = B^{-1}(y_0'', ..., y_4''), \quad (y_0'', ..., y_4'') = D(y_0', ..., y_4'),$$
$$(y_0', ..., y_4') = B(y_0, ..., y_4),$$

it will be found at once that we obtain

$$y_0''' = y_0 - y_1 - y_2 - y_3 - y_4, \quad ..., \quad y_4''' = -2y_0 - y_1 - y_2 - y_3 + 2y_4,$$

or, that $y_0''', ..., y_4'''$ are respectively identical with the $\eta_0, ..., \eta_4$ obtained above by interchange of x_1 and x_4. And this interchange is obtained (§ (7)) by projection from the node (14). Denoting it therefore by $p(14)$, we have

$$p(14) = B^{-1}DB,$$

where D, equivalent to projection from $(A_0 B)$, may be denoted by $p(A_0 B)$.

(d) It may also be remarked here that the transformation several times used, which we have denoted (§ (3)) by $(x') = \chi(x)$, when expressed by the variables $y_0, ..., y_4$, is equivalent to

$$y_0' = y_0, \quad y_1' = \epsilon^2 y_1, \quad y_2' = \epsilon y_2, \quad y_3' = \epsilon y_3, \quad y_4' = \epsilon^2 y_4.$$

Thus $\chi^3 = 1$. It may be verified also that $B^2 = p(A B_0)$, so that $B^4 = 1$, or $B^{-1} = B^3$, every projection being of period 2. Expressions for B in terms of projections are given below (§ (19)).

(14) **Explicit formulae for the rationalization of the Burkhardt primal.** We have seen (§ (6)) that there are 72.40 or 2880 ways in which the primal may be rationalized; we proceed to obtain the equations in one of these, shewing that *the prime sections of the primal correspond, in this way, to the quartic surfaces of a linear system in space of three dimensions.*

In the equation (I), $y_0^4 - y_0(y_1^3 + ... + y_4^3) + 3y_1 y_2 y_3 y_4 = 0$, denote $y_1/y_0, ..., y_4/y_0$ momentarily by a, b, c, d. Then we have

$$0 = 1 - a^3 - b^3 - c^3 - d^3 + 3abcd$$
$$= d^3 a^3 - d^3 - a^3 + 1 - (a^3 d^3 + b^3 + c^3 - 3adbc)$$
$$= (a^2 + a + 1)[(a - 1)(d^3 - 1) - (ad + b + c) BC],$$

where

$$B = (ad + \epsilon b + \epsilon^2 c)/(a - \epsilon), \quad C = (ad + \epsilon^2 b + \epsilon c)/(a - \epsilon^2);$$

from these equations for B, C, we obtain

$$b = a\left(d + \frac{\epsilon C - B}{1 - \epsilon}\right) + \frac{\epsilon B - C}{1 - \epsilon}, \quad c = a\left(d + \frac{\epsilon B - C}{1 - \epsilon}\right) - \frac{\epsilon^2 B - \epsilon^2 C}{1 - \epsilon},$$

whereby a birational transformation is defined, from a, b, c, d to a, B, C, d. Hence, as the equations for b, c lead to

$$b + c = a(2d - B - C) + \epsilon B + \epsilon^2 C,$$

the vanishing of $1 - a^3 - b^3 - c^3 - d^3 + 3abcd$, if $a^2 + a + 1$ does not vanish, leads to

$$a = [d^3 - 1 + BC(\epsilon B + \epsilon^2 C)]/[d^3 - 1 + BC(B + C - 3d)],$$

whereby the variable a is expressed rationally in terms of the three variables B, C, d, which, reversely, are rational functions of the four original variables a, b, c, d. This solves the problem.

We proceed to elaborate this result: First, instead of B, C use λ, μ, defined by $\lambda = B - d$, $\mu = C - d$; these are

$$\lambda = \epsilon^2(c + \epsilon^2 b + \epsilon^2 d)/(a - \epsilon), \quad \mu = \epsilon(c + \epsilon b + \epsilon d)/(a - \epsilon^2),$$

namely

$$\lambda = \epsilon^2(y_3 + \epsilon^2 y_2 + \epsilon^2 y_4)/(y_1 - \epsilon y_0), \quad \mu = \epsilon(y_3 + \epsilon y_2 + \epsilon y_4)/(y_1 - \epsilon^2 y_0),$$

and put ρ for d, so that $\rho = y_4/y_0$. Then, by means of the value found for a, the equation of the primal is solved rationally in terms of λ, μ, ρ, which are linear (fractional) functions of the original variables $y_0, y_1, ..., y_4$.

Thus we are led (cf. the formulation of §(6)) to consider the three planes

$$y_3 + \epsilon^2 y_2 + \epsilon^2 y_4 = 0 = y_1 - \epsilon y_0; \quad y_3 + \epsilon y_2 + \epsilon y_4 = 0 = y_1 - \epsilon^2 y_0;$$
$$y_0 - 0 - y_4.$$

By the general formulae of §(13), the equations of these planes in terms of the coordinates $x_1, ..., x_6$, are found to be, respectively,

$$x_1/1 = x_2/\epsilon^2 = x_5/\epsilon; \quad x_1/1 = x_3/\epsilon = x_6/\epsilon^2; \quad x_1/1 = x_2/\epsilon = x_3/\epsilon^2,$$

which are Jacobian planes, of which the second and third meet only in the point $(0, 0, 0, 1, -1, 0)$, or (45), while the third and first meet only in the point $(0, 0, 0, 1, 0, -1)$, or (46), and the first and second meet only in the point $(1, \epsilon^2, \epsilon, 1, \epsilon, \epsilon^2)$, or $(E_0 F)$; these three nodes lie in the Jacobian plane $x_1/1 = x_2/\epsilon^2 = x_3/\epsilon$, which contains also the six nodes (BC_0), (CA_0), (AB_0), $(F_0 D)$, $(D_0 E)$, (56). The rationalization is to be effected by means of the line, from the general point $(y_0, y_1, ..., y_4)$ of the primal, which meets these three skew planes; this line is given by the intersection of the three primes

$$\lambda = \epsilon^2 \frac{y_3 + \epsilon^2(y_2 + y_4)}{y_1 - \epsilon y_0}, \quad \mu = \epsilon \frac{y_3 + \epsilon(y_2 + y_4)}{y_1 - \epsilon^2 y_0}, \quad \rho = \frac{y_4}{y_0}.$$

Regarding λ, μ, ρ as the coordinates of the point in which this line meets an arbitrary solid, and denoting homogeneous variables in this solid by x, y, z, t, defined by $\rho = x/t$, $\lambda = y/t$, $\mu = z/t$, we may thus take, for the coordinates of the point of this solid which corresponds to $(y_0, y_1, ..., y_4)$, respectively,

$$x = y_4(y_1^2 + y_1 y_0 + y_0^2),$$
$$t = y_0(y_1^2 + y_1 y_0 + y_0^2),$$
$$y = \epsilon^2 y_0(y_1 - \epsilon^2 y_0)(y_3 + \epsilon^2 y_2 + \epsilon^2 y_4),$$
$$z = \epsilon y_0(y_1 - \epsilon y_0)(y_3 + \epsilon y_2 + \epsilon y_4);$$

and, using the value of a, or y_1/y_0, found above in terms of B, C, d, or B, C, ρ, and thence in terms of λ, μ, ρ, we compute the reverse expressions of $y_0, y_1, ..., y_4$ in terms of x, y, z, t, in the forms

$$y_0 = t[x(y^2 + yz + z^2) + yz(y + z) - t^3],$$
$$y_1 = -t[(1 - \epsilon)x^2(y - \epsilon^2 z) - \epsilon x(y - \epsilon^2 z)^2 - \epsilon yz(y + \epsilon z) + t^3],$$
$$y_2 = -y^2 z^2 - xyz(y + z) + (x + y + z)t^3,$$
$$y_3 = -\epsilon x^2(y - \epsilon^2 z)^2 + (1 - \epsilon)xyz(y - \epsilon^2 z) + y^2 z^2 - \epsilon(y + \epsilon z)t^3,$$
$$y_4 = x[x(y^2 + yz + z^2) + yz(y + z) - t^3].$$

The prime sections of the primal, which are given by an equation $\eta_0 y_0 + \eta_1 y_1 + ... + \eta_4 y_4 = 0$, wherein $\eta_0, \eta_1, ..., \eta_4$ are parametric constants, thus arise from a linear system of quartic surfaces in the threefold space in which x, y, z, t are coordinates.

These quartic surfaces, as we see from the manner in which y_2, y_3, y_4 have been formed, all pass through the common points of the two cubic surfaces arising in y_0 and y_1, namely

$$x(y^2 + yz + z^2) + yz(y + z) - t^3 = 0,$$

$$(1 - \epsilon) x^2(y - \epsilon^2 z) - \epsilon x(y - \epsilon^2 z)^2 - \epsilon yz(y + \epsilon z) + t^3 = 0,$$

but the addition of these two equations leads to

$$(1 - \epsilon)(y - \epsilon^2 z)(x + y)(x + z) = 0,$$

and this suggests, what we easily verify, that these two cubic surfaces have in common the nine lines, lying in threes in three planes, which are given by

(I) $y - \epsilon^2 z = 0$, $z^3 + t^3 = 0$; or $y = \epsilon^2 z = -\epsilon^k t$,

(II) $x + y = 0$, $y^3 + t^3 = 0$; or $x = -y = \epsilon^k t$, $(k = 0, 1, 2)$.

(III) $x + z = 0$, $z^3 + t^3 = 0$; or $x = -z = \epsilon^k t$.

If we put, for a moment,

$$X = -(\epsilon^2 - 1) x, \quad Y = (\epsilon^2 - 1) x - (y - \epsilon^2 z),$$

$$Z = y - z, \qquad T = (\epsilon^2 - 1) t,$$

the equations of the two cubic surfaces are found to be

$$X^3 + Y^3 + Z^3 + T^3 - 3(X + Y)(Y + \epsilon^2 Z)(Y + Z) = 0,$$

$$X^3 + Y^3 + Z^3 + T^3 = 0.$$

The latter surface has, for its twenty-seven lines, the joins of the points of the surface which lie on any edge of the tetrahedron $XYZT$ to the points lying in the opposite edge; and it is easy to verify that the nine lines common to the two surfaces consist of the six joining the two points $(X, Y, Z, T) = (0, 1, -1, 0)$, $(0, 1, -\epsilon, 0)$ in the edge YZ, to the three points of the opposite edge, together with the three which join $(1, -1, 0, 0)$, in XY, to the points in the opposite edge. Also, if the three lines named (I), for $k = 0, 1, 2$, be called a_1, b_1, c_1, and the three lines (II), for

$k = 0, 1, 2$, be called a_2, b_2, c_2, and the three lines (III), for
$k = 0, 1, 2$, be called a_3, b_3, c_3, it is found that the lines a_1, a_2, a_3,
c_1, c_2, c_3, b_1, b_2, b_3, in this order, form a closed polygon of nine sides.
Further, the former cubic surface has a double point at the point
$(X, Y, Z, T) = (1, -1, 0, 0)$, whereat the asymptotic cone consists of the two planes $X + Y = 0$, $X + Y - \epsilon Z = 0$. And it is easy
to see that the five quartic surfaces, in the space (x, y, z, t),
arising by putting $y_0 = 0$, $y_1 = 0, ..., y_4 = 0$ in the equations
above, all have nodes at the three points

$$(X, Y, Z, T) = (0, 1, -1, 0), \quad (0, 1, -\epsilon, 0), \quad (1, -1, 0, 0),$$

the asymptotic cone in each case breaking up into two planes, of
which one plane contains the three lines meeting at this node.

For a quartic surface to contain a line, five conditions must be
satisfied. For a quartic surface to contain the nine lines arising
here, it must contain the three points on the line of intersection of
any two of the three planes (I), (II), (III) in each of which three of
the lines lie, these being points of the cubic surfaces above; this
requires nine conditions. If the three lines in any one of these
planes had three different intersections, six further conditions
would be required for the surface to contain these three lines.
Then the $9 + 3.6$, or 27, conditions would leave eight linearly
independent quartic surfaces containing the nine given lines. On
examination we find that this remains true with the actual lines,
notwithstanding their intersections in threes, and that the
general quartic surface containing the nine lines has an equation

$$p(x^3 - t^3)(y - \epsilon^2 z) + q(z^3 + t^3)(x + y) + r(y^3 + t^3)(x + z)$$

$$+ \eta_0 y_0 + \eta_1 y_1 + ... + \eta_4 y_4 = 0,$$

where $p, q, r, \eta_0, \eta_1, ..., \eta_4$ are arbitrary, and $y_0, y_1, ..., y_4$ are the
quartic polynomials in x, y, z, t put down above. If now we add
the conditions that the quartic surfaces shall have double points
at each of the points of concurrence of threes of the lines, for
which $(x, y, z, t) = (1, 0, 0, 0)$, $(0, 0, 1, 0)$, $(0, 1, 0, 0)$, we find that
p, q, r must all vanish.

Thus the quartic surface found above is the most general surface containing the nine common lines of the two cubic surfaces which has a double point at each of the three points of concurrence of threes of the lines, and the Burkhardt primal can be defined by the fact that its prime sections correspond to the system of such quartic surfaces.

In connexion with this result it may be remarked that the equation (I), or $y_0^4 - y_0(y_1^3 + \ldots + y_4^3) + 3y_1 y_2 y_3 y_4 = 0$, is satisfied identically by putting

$$y_0 = 3xyzt, \qquad y_1 = x(y^3 - z^3 - t^3), \quad y_2 = y(z^3 - x^3 - t^3),$$

$$y_3 = z(x^3 - y^3 - t^3), \quad y_4 = t(x^3 + y^3 + z^3);$$

but, reversely, these equations give six points (x, y, z, t) corresponding to the point $(y_0, y_1, \ldots, y_4)$. While, if we take all the points of the primal which lie on the tangent prime at a particular point, P, the sets of six points of the space (x, y, z, t) which correspond to the points of this prime section, describe the Weddle quartic surface whose nodes are the six points corresponding to the point P. This remark is due to Coble (*loc. cit.*); it furnishes an example of an involution, in space of three dimensions, which is rational (the reader may compare the author's *Principles of Geometry*, Vol. VI, p. 137).

As has been stated, the fact that the Burkhardt primal is rational was recognized by Dr J. A. Todd (*loc. cit.*), with a masterly argument differing from that here employed, which, however, does not readily lead to the reverse equations.

(15)   **The equation of the Burkhardt primal referred to the prime faces of a Jordan pentahedron.** From the equations of the faces of a Jordan pentahedron, which are linear functions of $x_1, \ldots, x_6$, we may express these latter in terms of the former. We take the pentahedron $\{A\}$, whose faces are given by the vanishing of the forms

$$\xi_1 = [14.25.36], \quad \xi_2 = [16.35.24], \quad \xi_3 = [13.26.45],$$

$$\xi_4 = [12.34.56], \quad \xi_5 = [15.46.23],$$

the equations of transformation, and their reverses, being

$$(0, \xi_1, \xi_2, \ldots, \xi_5) = \begin{pmatrix} 1 & 1 & 1 & 1 & 1 & 1 \\ 1 & \epsilon & \epsilon^2 & 1 & \epsilon & \epsilon^2 \\ 1 & \epsilon^2 & \epsilon & \epsilon^2 & \epsilon & 1 \\ 1 & \epsilon & 1 & \epsilon^2 & \epsilon^2 & \epsilon \\ 1 & 1 & \epsilon & \epsilon & \epsilon^2 & \epsilon^2 \\ 1 & \epsilon^2 & \epsilon^2 & \epsilon & 1 & \epsilon \end{pmatrix} (x_1, x_2, \ldots, x_6),$$

$$6(x_1, x_2, \ldots, x_6) = \begin{pmatrix} 1 & 1 & 1 & 1 & 1 & 1 \\ 1 & \epsilon^2 & \epsilon & \epsilon^2 & 1 & \epsilon \\ 1 & \epsilon & \epsilon^2 & 1 & \epsilon^2 & \epsilon \\ 1 & 1 & \epsilon & \epsilon & \epsilon^2 & \epsilon^2 \\ 1 & \epsilon^2 & \epsilon^2 & \epsilon & \epsilon & 1 \\ 1 & \epsilon & 1 & \epsilon^2 & \epsilon & \epsilon^2 \end{pmatrix} (0, \xi_1, \ldots, \xi_5);$$

by means of these, the primes of any one of the pentahedra can be expressed in terms of $\xi_1, \ldots, \xi_5$. In terms of $\xi_1, \ldots, \xi_5$, the equation of the Burkhardt primal is found to be

$$\xi_1^2 \xi_2^2 + \xi_2^2 \xi_4^2 + \xi_4^2 \xi_5^2 + \xi_5^2 \xi_3^2 + \xi_3^2 \xi_1^2$$
$$+ \epsilon(\xi_1^2 \xi_4^2 + \xi_2^2 \xi_5^2 + \xi_4^2 \xi_3^2 + \xi_5^2 \xi_1^2 + \xi_3^2 \xi_2^2) = 0.$$

This equation contains only squares of the coordinates, and puts in evidence the fact (§(7)) that any one of the angular points of the pentahedron $\{A\}$ is a node, say O, and a node of such a character that, if P be any point of the primal, and OP meet the primal again in P', then the locus of the harmonic point $(P, P')/O$, is the prime face of the pentahedron opposite to O. By §(7) we presume, without making the computation, that the equation of the primal, when referred to any other of the pentahedra, likewise contains only squares of the coordinates.

If in the equation we have found we put

$$x = \epsilon^2 \xi_1, \quad y = \epsilon^2 \xi_2, \quad z = \xi_3, \quad t = \xi_4, \quad u = \xi_5,$$

the equation becomes

$$y^2 z^2 + x^2 t^2 + \epsilon^2(z^2 x^2 + y^2 t^2) + \epsilon(x^2 y^2 + z^2 t^2)$$
$$+ u^2(x^2 + y^2 + z^2 + t^2) = 0, \quad \text{(II)}$$

this shews that the twelve nodes in the face $u = 0$ form three tetrahedra with angular points given by the rows of the matrices

$$\begin{pmatrix} 1,0,0,0 \\ 0,1,0,0 \\ 0,0,1,0 \\ 0,0,0,1 \end{pmatrix}, \quad \begin{pmatrix} 1,-1,-1,&1 \\ -1,&1,-1,&1 \\ -1,-1,&1,&1 \\ -1,-1,-1,-1 \end{pmatrix}, \quad \begin{pmatrix} -1,&1,&1,&1 \\ 1,-1,&1,&1 \\ 1,&1,-1,&1 \\ 1,&1,&1,-1 \end{pmatrix},$$

for, as is easy to see, these are the nodes of any quartic surface, in the space (x, y, z, t), with equation

$$a(y^2z^2 + x^2t^2) + b(z^2x^2 + y^2t^2) + c(x^2y^2 + z^2t^2) = 0,$$

provided $a + b + c = 0$; by a similar argument the same is true for the nodes in any face of the pentahedron. Thus there are sixteen κ-lines containing the nodes of any face (§ (5)).

We have presumed that a similar form of equation holds for any pentahedron. This is actually so; see § (22) below. In particular, taking the pentahedron $\{AB\}$, and putting

$$X = (\epsilon^2 - \epsilon)[25], \quad Y = [A_0B], \quad Z = [AB_0],$$
$$T = (\epsilon^2 - \epsilon)[36], \quad U = (\epsilon^2 - \epsilon)[14],$$

it may be verified that the relation connecting X, Y, Z, T, U is precisely the same as that above connecting x, y, z, t, u. The linear expressions for X, Y, Z, T, U in terms of x, y, z, t, u thus give a transformation of the equation into itself. With

$$P = \epsilon y + z, \quad Q = \epsilon y - z, \quad A = \epsilon^2(t + u), \quad B = \epsilon^2(t - u),$$

these equations, save for a common factor, are

$$X = \epsilon^2(Q + B), \quad T = Q - B, \quad Z = -\epsilon(P + A),$$
$$U = \epsilon(P - A), \quad Y = 2x.$$

The equation of the primal in terms of x, y, z, t, u may conveniently be used to give more precision to the remarks above (§ (9)) on the relations between a Steiner tetrahedron and a Jordan prime. Let the angular points of the polyhedron considered which are opposite to the faces $u = 0$ and $t = 0$ be denoted respectively by U and T. The asymptotic cones of the primal at these angular points are given by the respective equations

$$x^2 + y^2 + z^2 + t^2 = 0, \quad x^2 + \epsilon^2 y^2 + \epsilon z^2 + u^2 = 0.$$

Four primes can be drawn through the line UT to touch both these cones, namely the four primes given by $x \pm \epsilon^2 y \pm \epsilon z = 0$. Consider one of these, say $x + \epsilon^2 y + \epsilon z = 0$; let its line of contact with the cone (U) be called l, and its line of contact with the cone (T) be called m; these lines do not meet. Through the line l there pass two Jacobian planes, lying both on the cone (U), and on the primal, as we easily verify, namely

$$x + \epsilon^2 y + \epsilon z = 0 = t + \epsilon y - \epsilon^2 z; \quad x + \epsilon^2 y + \epsilon z = 0 = -t + \epsilon y - \epsilon^2 z;$$

likewise, through the line m there pass two Jacobian planes lying both on the cone (T), and on the primal, namely

$$x + \epsilon^2 y + \epsilon z = 0 = u + y - z; \quad x + \epsilon^2 y + \epsilon z = 0 = -u + y - z.$$

These four planes are the faces of a Steiner tetrahedron. In fact, the line l has for equations $x/1 = y/\epsilon^2 = z/\epsilon = t/0$, and the two Jacobian planes passing through this line, when expressed in terms of the coordinates $x_1, \ldots, x_6$, are found to be

$$x_1/1 = x_3/\epsilon = x_6/\epsilon^2$$

and $x_2/1 = x_4/\epsilon = x_5/\epsilon^2$, the prime $x + \epsilon^2 y + \epsilon z = 0$ being

$$x_1 + x_3 + x_6 = 0.$$

These equations identify the line l as a κ-line, containing, besides the point U, which is (AF_0), the two other nodes

$$x = y/\epsilon^2 = z/\epsilon = t/0 = u/\pm 1;$$

in regard to these two nodes, the point in which the line meets the prime $u = 0$ is the harmonic of U. Likewise, the two planes through the line m, whose equations are $x = y = z$; $u = 0$, are Jacobian planes, forming the other two faces of the Steiner tetrahedron in the solid $x_1 + x_3 + x_6 = 0$. The Steiner tetrahedron has four edges besides l and m, on each of which are three other nodes; these nodes can be shewn to be the intersections of the

4-2

edge considered with the three prime faces of the Jordan pentahedron which contain the edge-line (these, with $u = 0$, $t = 0$,
making the five faces of the pentahedron). Each of these prime
faces, by its intersection with the Steiner solid, thus gives a plane
containing six nodes, one on each edge of the Steiner tetrahedron.
This plane is therefore one of the 540 cross-planes considered in
§ (9).

Further, it can be verified that the tangent prime of the
Burkhardt primal at any point of the line l, or at any point of the
line m, is the Steiner solid $x + \epsilon^2 y + \epsilon z = 0$. Thus this prime
touches the primal at every ordinary point of every edge of the
Steiner tetrahedron. This we may also verify simply from the
equation of the primal in the coordinates $y_0, y_1, ..., y_4$. More
generally, the tangent prime of the primal, at any ordinary point
of one of the cubic curves which pass through the nine nodes of a
Jacobian plane, touches the primal at every point of the curve.
The prime $x + \epsilon^2 y + \epsilon z = 0$ also touches the cones (U), (T) along
the lines l, m, as we have said.

We have considered only one of the common tangent primes of
the cones (U), (T) which can be drawn through the line UT; and
there are four such primes. Moreover, the pentahedron has ten
edges such as TU. Thus, from the single pentahedron we may
construct forty Steiner spaces.

The equation of the Burkhardt primal in terms of squares of
$\xi_1, ..., \xi_5$, considered here, shews that the primal contains an involution of sets of sixteen points, represented by a quadric. We
return to this later (§ (22)).

(16) **The thirty-six double sixes of Jordan pentahedra,
and the associated quadrics.** We have seen (§ (5)) that, taking
any particular pentahedron, there are sixteen others of which no
one has an angular point common with the pentahedron taken;
we may speak of two pentahedra which have no common angular
point as being *skew* with one another. Besides the nodes of the
pentahedron taken, there are, in each face ϖ, of this pentahedron,

eight nodes, each arising as an angular point of a pentahedron, having the node P, opposite to ϖ in the pentahedron taken, as an angular point. With each of these eight points as angular points, there are two other pentahedra, leading to sixteen pentahedra for the face ϖ. The sixteen pentahedra similarly obtainable by considering any other face of the pentahedron taken, coincide, in aggregate, with those arising for the face ϖ; these are the sixteen which are skew with the pentahedron taken.

In this way, or by direct specification of the angular points of each pentahedron, we see that the sixteen pentahedra, which are skew with the pentahedron $\{A\}$, consist of the six $\{A_0\}$, $\{B\}$, $\{C\}$, $\{D\}$, $\{E\}$, $\{F\}$ and the ten $\{PQ\}$, where P, Q are two of B, C, D, E, F; with a similar statement for the pentahedra skew with $\{A_0\}$. Also we see that the sixteen pentahedra which are skew with $\{AB\}$ consist of the eight $\{AP\}$, $\{BP\}$, where P is one of C, D, E, F, together with the eight given by $\{P\}$, $\{P_0\}$. Thus also $\{A\}$ has an angular point common with the five $\{B_0\}$, $\{C_0\}$, $\{D_0\}$, $\{E_0\}$, $\{F_0\}$, and with any one of the five $\{AP\}$, where P is one of B, C, D, E, F, with a similar statement for $\{A_0\}$; and $\{AB\}$ has an angular point common with $\{A\}$, $\{A_0\}$, $\{B\}$, $\{B_0\}$, or with any one of $\{PQ\}$, where P, Q are two of C, D, E, F.

Bearing these results in mind, consider now the three sets, each of twelve pentahedra, which we arrange in two rows of six pentahedra:

$$
\begin{array}{llllll}
\{A\}, & \{B\}, & \{C\}, & \{D\}, & \{E\}, & \{F\} \\
\{A_0\}, & \{B_0\}, & \{C_0\}, & \{D_0\}, & \{E_0\}, & \{F_0\}
\end{array} \Bigg\} \text{ (I),}
$$

$$
\begin{array}{llllll}
\{A\}, & \{A_0\}, & \{BC\}, & \{BD\}, & \{BE\}, & \{BF\} \\
\{B\}, & \{B_0\}, & \{AC\}, & \{AD\}, & \{AE\}, & \{AF\}
\end{array} \Bigg\} \text{ (II),}
$$

$$
\begin{array}{llllll}
\{A\}, & \{B\}, & \{C\}, & \{EF\}, & \{FD\}, & \{DE\} \\
\{BC\}, & \{CA\}, & \{AB\}, & \{D_0\}, & \{E_0\}, & \{F_0\}
\end{array} \Bigg\} \text{ (III).}
$$

From the examination just made, it appears that in any one of these *double sixes*, any pentahedron is skew with the others in the same row, and with that in the other row which is in the same

column, but has an angular point in common with the five
pentahedra in the other row which are not in the same column.
There are fifteen double sixes of the type (II), obtained by
replacing A, B by another pair, and there are twenty double sixes
of the type (III), obtained by replacing A, B, C by another triad.
Thus in all there are thirty-six double sixes.* It is easy to see that
the thirty angular points of the pentahedra which occur in one
row of a double six are the same, in different order, as those
arising in the other row. The thirty-six double sixes contain in all
seventy-two sets of six pentahedra, of which every two of a set
are skew to one another. It can be seen that no two of these
seventy-two sets have the same set of *four* pentahedra common to
both; thus the thirty nodes arising in a double six are all determined when a proper set of twenty nodes is given.

To each double six of pentahedra there corresponds a quadric,
which contains the thirty nodes arising in the double six, and
touches the thirty polar primes which are the faces of the pentahedra; the quadric is both inscribed and circumscribed to every
one of the twelve pentahedra of the double six. For the three
representative double sixes (I), (II), (III) put down above, the
proper quadrics are respectively

$$x_1^2 + x_2^2 + \ldots + x_6^2 = 0,$$

$$x_1 x_4 + x_2 x_5 + x_3 x_6 = 0, \quad (x_1 + x_2 + \ldots + x_6 = 0)$$

$$x_2 x_3 + x_3 x_1 + x_1 x_2 + \epsilon(x_5 x_6 + x_6 x_4 + x_4 x_5) = 0.$$

That the first quadric contains all the thirty nodes of the
double six (I) is easy to verify; and taking, for example, the node
$(A_0 B)$, in the polar prime $[A_0 B] = 0$ of the pentahedron $\{A\}$, the
quadric touches this prime at this point (which is an angular
point of $\{A_0\}$). Similarly for each of the six columns of the double

---

* The existence of such double sixes, for the lines of a cubic surface,
was remarked by Schläfli, *Quart. J. Math.* II (1858). The double sixes of
Jordan pentahedra are noticed by Burkhardt, *Math. Ann.* XXXVIII (1890),
p. 197, who gives the pentahedra of one of these, expressed in terms of
$y_0, y_1, \ldots, y_4$.

six. We may form generally the condition that a prime

$$u_1 x_1 + \dots + u_6 x_6 = 0$$

should touch the quadric. As the same prime is represented by this equation if all of $u_1, \dots, u_6$ are increased by the same arbitrary quantity, this condition of tangency is a function of the differences of $u_1, \dots, u_6$; but we may follow the plan tacitly adopted in the notation for the Jordan primes, of introducing the condition $u_1 + \dots + u_6 = 0$. The *tangential equation* so obtained, which, in general form, is

$$u_1^2 + u_2^2 + \dots + u_6^2 - \tfrac{1}{6}(u_1 + \dots + u_6)^2 = 0,$$

thus reduces to

$$u_1^2 + u_2^2 + \dots + u_6^2 = 0, \quad (u_1 + u_2 + \dots + u_6 = 0).$$

From this it is at once recognized that the quadric touches the thirty prime faces of the pentahedra belonging to the double six. Also, for each of these twelve pentahedra, the point of contact, with any prime face, is the angular point, lying on this face, of the other pentahedron in the same column of the double six.

Later, we shall find it interesting (§ (22)) to form the equation of the quadric in terms of the prime faces of one of its component pentahedra.

We can similarly verify directly that the thirty nodes occurring in the double six (II) lie, as stated, on the quadric

$$x_1 x_4 + x_2 x_5 + x_3 x_6 = 0,$$

and that this touches the thirty prime faces of the pentahedra, the rule for the point of contact being precisely as in the former case. Also, though we omit the algebra, which is given below for the less easy case (III), we can verify that the tangential equation of this quadric is

$$u_1 u_4 + u_2 u_5 + u_3 u_6 - \tfrac{1}{12}(u_1 + \dots + u_6)^2 = 0,$$

or $\quad u_1 u_4 + u_2 u_5 + u_3 u_6 = 0, \quad (u_1 + u_2 + \dots + u_6 = 0).$

We can obtain the equation $x_1 x_4 + x_2 x_5 + x_3 x_6 = 0$ by trial. But we can also obtain the equation from the former,

$$x_1^2 + \dots + x_6^2 = 0,$$

by remarking that the thirty nodes occurring in the double six (II) consist of the eighteen nodes arising from $\{A\}$, $\{A_0\}$, $\{B\}$, $\{B_0\}$ (wherein each of (AB_0), (A_0B) arises twice), together with the fifteen nodes (pq) other than (14), (25), (36), which are the other twelve. That these twelve lie on the quadric $x_1x_4 + x_2x_5 + x_3x_6 = 0$ is clear. The other eighteen consist of two sets of nine given by

$$(A_0B);\quad (AC_0),\quad (AD_0),\quad (AE_0),\quad (AF_0);$$
$$(B_0C),\quad (B_0D),\quad (B_0E),\quad (B_0F);$$
$$(AB_0);\quad (A_0C),\quad (A_0D),\quad (A_0E),\quad (A_0F);$$
$$(BC_0),\quad (BD_0),\quad (BE_0),\quad (BF_0),$$

all of which lie on $x_1^2 + \ldots + x_6^2 = 0$; but the first set lie upon $[A_0B] = 0$, whose equation, put into the form

$$(x_1+x_4)/1 = (x_2+x_5)/\epsilon = (x_3+x_6)/\epsilon^2,$$

shews that these lie on the degenerate cone

$$(x_1+x_4)^2 + (x_2+x_5)^2 + (x_3+x_6)^2 = 0;$$

while the second set, lying on $[AB_0] = 0$, similarly also lie on this cone. The equation of the quadric belonging to the double six (II) is thus clear from the identity

$$2(x_1x_4 + x_2x_5 + x_3x_6)$$
$$= (x_1+x_4)^2 + (x_2+x_5)^2 + (x_3+x_6)^2 - (x_1^2 + x_2^2 + \ldots + x_6^2).$$

To illustrate the general rule given above, for the quadric $x_1x_4 + x_2x_5 + x_3x_6 = 0$, it is easy to see that the tangent prime of this quadric at the point (A_0B), which is an angular point of $\{B\}$ lying in the polar prime $[A_0B] = 0$, opposite to the angular point (AB_0) of $\{A\}$, is $[A_0B] = 0$. Or, considering the pentahedron $\{BC\}$, the prime face opposite to the angular point (BC_0), namely $[B_0C] = 0$, or $[15.26.34] = 0$, is the tangent prime of the quadric at $(16.24.35)$, or (AC_0), which is an angular point of $\{AC\}$. Or, still again, the prime face of $\{BC\}$ opposite to the angular point (15), namely $[15] = 0$, is the tangent prime of the quadric at the node (24), which is also an angular point of $\{AC\}$.

Similarly, for the double six of type (III), for example, the angular points of the pentahedron $\{EF\}$ are

$$(1, \epsilon, \epsilon^2, 1, \epsilon^2, \epsilon), \quad (1, \epsilon^2, \epsilon, 1, \epsilon, \epsilon^2), \quad (14), \quad (26), \quad (35),$$

which are easily seen to lie on the quadric whose equation has been given; also the prime face of $\{EF\}$ opposite to (EF_0), namely $[E_0 F] = 0$, is $x_1 + \epsilon^2 x_2 + \epsilon x_3 + x_4 + \epsilon x_5 + \epsilon^2 x_6 = 0$; and this is the tangent prime of the quadric at the point $(1, \epsilon^2, \epsilon, \epsilon^2, 1, \epsilon)$, or $(15.36.24)$, or $(D_0 E)$, which is an angular point of the pentahedron $\{D_0\}$, occurring in the double six in the same column as $\{EF\}$. Or again, the prime face of $\{EF\}$ opposite to (14), is $x_1 - x_4 = 0$, which is the tangent prime of the quadric at (CD_0), also an angular point of $\{D_0\}$.

That the quadric touches the primes of the pentahedra of the double six (III), is similarly verifiable from the tangential equation of the quadric. In complete form this equation is

$$12[u_2 u_3 + u_3 u_1 + u_1 u_2 + \epsilon^2 (u_5 u_6 + u_6 u_4 + u_4 u_5)]$$
$$+ [(\epsilon^2 - 3) \sigma + (1 - 3\epsilon^2) \tau] [\sigma + \tau] = 0,$$

where σ denotes $u_1 + u_2 + u_3$, and τ denotes $u_4 + u_5 + u_6$. If we use $u_1 + \ldots + u_6 = 0$, or $\sigma + \tau = 0$, this reduces to

$$u_2 u_3 + u_3 u_1 + u_1 u_2 + \epsilon^2 (u_5 u_6 + u_6 u_4 + u_4 u_5) = 0.$$

We may give indications of the algebra by which this tangential equation is found: The tangent prime of a quadric

$$f(x_1, x_2, \ldots, x_6) = 0, \quad \text{or} \quad f(x_1, \ldots, x_5, -x_1 - x_2 - \ldots - x_5) = 0,$$

at the point $(\xi_1, \ldots, \xi_6)$, where $\xi_1 + \ldots + \xi_6 = 0$, if F denote $f(\xi_1, \ldots, \xi_6)$, is given by

$$x_1 \left(\frac{\partial F}{\partial \xi_1} - \frac{\partial F}{\partial \xi_6} \right) + \ldots + x_5 \left(\frac{\partial F}{\partial \xi_5} - \frac{\partial F}{\partial \xi_6} \right) = 0,$$

and this is

$$x_1 \partial F / \partial \xi_1 + \ldots + x_6 \partial F / \partial \xi_6 = 0, \quad \text{or, say,} \quad x_1 F_1 + \ldots + x_6 F_6 = 0.$$

To identify this with $u_1 x_1 + \ldots + u_6 x_6 = 0$, we require

$$F_1 / (u_1 + \theta) = F_2 / (u_2 + \theta) = \ldots = F_6 / (u_6 + \theta), \quad = \lambda / \theta, \text{ say,}$$

where θ is arbitrary; and we are to eliminate $\xi_1, ..., \xi_6$ by means of $F = 0$. For the case where

$$F = \xi_2\xi_3 + \xi_3\xi_1 + \xi_1\xi_2 + \epsilon(\xi_5\xi_6 + \xi_6\xi_4 + \xi_4\xi_5)$$

the equations give

$$\xi_2 + \xi_3 = \frac{\lambda}{\theta}(u_1 + \theta), \quad \xi_3 + \xi_1 = \frac{\lambda}{\theta}(u_2 + \theta), \quad \xi_1 + \xi_2 = \frac{\lambda}{\theta}(u_3 + \theta),$$

$$\xi_5 + \xi_6 = \epsilon^2\frac{\lambda}{\theta}(u_4 + \theta), \quad \xi_6 + \xi_4 = \epsilon^2\frac{\lambda}{\theta}(u_5 + \theta), \quad \xi_4 + \xi_5 = \epsilon^2\frac{\lambda}{\theta}(u_6 + \theta),$$

so that, with $\sigma = u_1 + u_2 + u_3$, $\tau = u_4 + u_5 + u_6$, because

$$\xi_1 + \xi_2 + ... + \xi_6 = 0,$$

we have $\sigma + \epsilon^2\tau = 3\epsilon\theta$; hence, putting M for $\sigma + \epsilon^2\tau$, we have such equations as

$$2\xi_1 = \frac{\lambda}{\theta}(\sigma - 2u_1 + \tfrac{1}{3}\epsilon^2 M), \quad 2\xi_4 = \epsilon^2\frac{\lambda}{\theta}(\tau - 2u_4 + \tfrac{1}{3}\epsilon^2 M);$$

if these values of $\xi_1, ..., \xi_6$ be substituted in $F = 0$, the result is the equation we have stated.

It follows from the geometrical interpretation that any linear transformation of the coordinates which leaves the Burkhardt primal unchanged must change the set of six skew pentahedra which form one of the seventy-two rows of the double sixes, into another such set (or into itself). Such transformation must therefore change the thirty nodes arising for this set into another such system, and change the quadric containing these into another such quadric. In particular the projections of the primal into itself explained in § (7) have this property; and a projection from one of the thirty nodes of a system changes this system into another such system, and the quadric containing the first system into the proper quadric of the second. We can exemplify this with precise formulae which will be found to be of interest below (Appendix, Note 2). Let the rows of six pentahedra occurring in the double sixes above called (I), (II), (III) be denoted respectively by

$$\begin{pmatrix} \alpha \\ \alpha_0 \end{pmatrix}, \quad \begin{pmatrix} \nu_{AB} \\ \nu_{BA} \end{pmatrix}, \quad \begin{pmatrix} \lambda_{ABC} \\ \mu_{ABC} \end{pmatrix},$$

the order of the pentahedra in a row not being regarded. Let projection from a node, say (AB_0), be denoted by $p(AB_0)$. Then it is found, by the rules given in § (7) (see below), that

$$p(A_0 B) \cdot \nu_{BA} = p(AB_0) \cdot \nu_{AB} = \alpha; \qquad p(AB_0) \cdot \nu_{AC} = \lambda_{ABC};$$

$$p(A_0 B) \cdot \nu_{AB} = p(AB_0) \cdot \nu_{BA} = \alpha_0; \qquad p(AB_0) \cdot \nu_{CA} = \mu_{ABC}.$$

These relations shew how to obtain, from the single set α, by use of the projections, all the remaining seventy-one sets of six skew pentahedra, every one of these sets being of one of the types α, α_0, ν_{AB}, ν_{BA}, λ_{ABC}, μ_{ABC}. We shall shew below (§ (20)) how to permute the six pentahedra $\{A\}, \{B\}, ..., \{F\}$ of the set α among themselves by a group of 360 projections. We can thus generate a group of 72.360, or $2^3.3^4.40$, or 25920 projections.

If the quadrics associated with the three typical double sixes (I), (II), (III) be denoted respectively by Q, Q_{AB}, Q_{ABC}, the relations put down shew that

$$p(AB_0) \cdot Q_{AB} = Q, \quad p(AB_0) \cdot Q_{AC} = Q_{ABC}.$$

Particular consequences are that the quadric Q_{AB} is unaltered by the transformation $(x') = \chi(x)$, employed in § (3), as may easily be verified directly; and that the quadric Q_{ABC} is unaltered by the transformation $(x') = \psi(x)$ given by $p(AB_0) p(B_0 C)$, whose expression in $x_1, ..., x_6$ is

$$x_1' = \epsilon x_4 + \epsilon x_5 + x_6, \quad -x_4' = \epsilon^2 x_1 + x_2 + \epsilon^2 x_3,$$

$$x_2' = x_4 + \epsilon x_5 + \epsilon x_6, \quad -x_5' = \epsilon^2 x_1 + \epsilon^2 x_2 + x_3,$$

$$x_3' = \epsilon x_4 + x_5 + \epsilon x_6, \quad -x_6' = x_1 + \epsilon^2 x_2 + \epsilon^2 x_3.$$

It may be desirable to illustrate the process by which the equations above are obtained by taking some cases. It is to be borne in mind, see § (7), that nodes (PQ_0), (QR_0), (RP_0) lie on a κ-line, and that nodes (PQ_0), (RS_0), (ij) lie on a κ-line if the syntheme symbols of (PQ_0), (RS_0) are interchanged by the transposition of the numbers i and j; and then that, if (L), (M) be two angular points of the same pentahedron $p(L) \cdot (M) = (M)$; while, if L, M, N be three nodes of a κ-line, $p(L) \cdot (M) = (N)$.

Also that $p(L).(L) = (L)$. As the projection $p(AB_0)$ is of period 2, one of the equations put down is

$$p(AB_0).\alpha = \nu_{AB}.$$

Now the left side consists of the elements

$$p(AB_0).\{A\}, \quad p(AB_0).\{B\}, \quad p(AB_0).\{C\}, \quad ..., \quad p(AB_0).\{F\}.$$

The geometrical interpretation shews that $p(AB_0).\{A\} = \{A\}$. The element $p(AB_0).\{B\}$ consists of five terms such as

$$p(AB_0).(BP_0),$$

where P is one of $A, C, ..., F$, and this term is (PA_0) except when P is A, when it is (BA_0); thus $p(AB_0).\{B\} = \{A_0\}$. The element $p(AB_0).\{C\}$ consists of five terms $p(AB_0).(CP_0)$, where P is one of A, B, D, E, F; of these

$$p(AB_0).(CA_0) = (BC_0), \quad p(AB_0).(CB_0) = (CB_0),$$
$$p(AB_0).(CD_0) = (26), \quad p(AB_0).(CE_0) = (34),$$
$$p(AB_0).(CF_0) = (15),$$

as we see at once by consulting the scheme of synthemes in § (1); thus $p(AB_0).\{C\} = \{BC\}$. In the same way we find

$$p(AB_0).\{P\} = (BP)$$

for $P = D, E, F$. On the whole then $p(AB_0).\alpha = \nu_{AB}$.

Consider next the equation

$$p(AB_0).\nu_{AC} = \lambda_{ABC}.$$

The left side of this consists of the elements

$$p(AB_0).\{A\}, \qquad p(AB_0).\{A_0\}, \qquad p(AB_0).\{CB\},$$
$$p(AB_0).\{CD\}, \qquad p(AB_0).\{CE\}, \qquad p(AB_0).\{CF\}.$$

Of these, $p(AB_0).\{A\} = \{A\}$; $p(AB_0)\{A_0\}$ consists of terms $p(AB_0).(A_0P)$, equal to (BP_0), so that $p(AB_0)\{A_0\} = \{B\}$, as we have already seen; $p(AB_0).\{CB\}$ is already found above to be $\{C\}$; $p(AB_0).\{CD\}$ consists of terms $p(AB_0).(CD_0), p(AB_0).(C_0D)$, $p(AB_0).(14), p(AB_0).(23), p(AB_0).(56)$, which are respectively $(26), (35), (14), (EF_0), (E_0F)$. Thus we have $p(AB_0).\{CD\} = \{EF\}$. So we find $p(AB_0).\{CE\} = \{FD\}$, and $p(AB_0).\{CF\} = \{DE\}$. On the whole then

$$p(AB_0) \cdot \nu_{AC} = \{A\}, \{B\}, \{C\}, \{EF\}, \{FD\}, \{DE\},$$

as stated. The remaining equations can be similarly verified.

(17) **The linear transformations of the Burkhardt primal into itself.** We have shewn that the equation of the primal is reducible to the form

$$y_0^4 - y_0(y_1^3 + y_2^3 + y_3^3 + y_4^3) + 3y_1 y_2 y_3 y_4 = 0,$$

by taking $y_0 = 0$ to be a Steiner solid meeting the primal in the four planes $y_0 = y_1 = 0, \ldots, y_0 = y_4 = 0$. With the values taken in § (13) these planes are respectively

$$x_1/1 = x_2/\epsilon^2 = x_3/\epsilon; \quad x_4/1 = x_5/\epsilon^2 = x_6/\epsilon;$$
$$x_4/1 = x_5/\epsilon = x_6/\epsilon^2; \quad x_1/1 = x_2/\epsilon = x_3/\epsilon^2.$$

We are concerned with the linear transformations into itself of which the primal is capable; and these may be expressed by any set of coordinates. Such a set as $y_0, y_1, \ldots, y_4$ was used by Burkhardt, arising in his theory as theta functions of two variables, linearly transformed among themselves when the thirds of the periods of the theta functions undergo a linear transformation. From this point of view it appears that the linear transformations of $y_0, y_1, \ldots, y_4$ with which we are concerned can be built up *by the combinations of four such transformations*.[*] It is proper to cite these transformations with the names given to them by Burkhardt omitting, however, constant factors common to all of $y_0, y_1, \ldots, y_4$, and replacing his S_2 by S. They are

	B	C	D	S
y_0'	$y_0 - y_1$	y_0	y_0	y_0
y_1'	$-2y_0 - y_1$	y_1	y_2	$\epsilon^2 y_1$
y_2'	$y_2 + y_3 + y_4$	y_4	y_1	y_2
y_3'	$y_2 + \epsilon y_3 + \epsilon^2 y_4$	y_2	y_3	$\epsilon^2 y_3$
y_4'	$y_2 + \epsilon^2 y_3 + \epsilon y_4$	y_3	y_4	$\epsilon^2 y_4$

[*] The reader interested in the linear transformation of the periods of theta functions may consult, for instance, the writer's *Abel's Theorem and the Theory of the Theta Functions* (Cambridge, 1897), Chap. xviii, and pp. 549, 669 ff., where references to the literature are given.

These transformations are subject to various relations among themselves (see below, §(19)); in particular we mention at once

$$B^4 = C^3 = D^2 = S^3 = 1, \quad B^2 = C^2 D C D C^2 D,$$
$$(CS)^3 = (SC)^3 = DS^2D.$$

The transformations C, D, S are transformations of the faces of the Steiner tetrahedron lying in $y_0 = 0$, or $x_1 + x_2 + x_3 = 0$, among themselves. But B transforms this Steiner solid into the Steiner solid $y_0 - y_1 = 0$, or $x_2 - \epsilon x_3 = 0$, which as already remarked, §(9), meets the original Steiner solid in the Jacobian plane

$$x_1/1 = x_2/\epsilon = x_3/\epsilon^2.$$

It is at once evident that C, D, S leave the primal unaltered. That the same is true of B follows because

$$y_0' = y_0 - y_1, \quad y_1' = -2y_0 - y_1,$$

lead to

$$y_0' - y_1' = 3y_0, \quad y_0'^2 + y_0'y_1' + y_1'^2 = 3(y_0^2 + y_0 y_1 + y_1^2),$$
$$y_0'^4 - y_0'y_1'^3 = 9(y_0^4 - y_0 y_1^3),$$

while
$$-y_0'(y_2'^3 + y_3'^3 + y_4'^3) + 3y_1'y_2'y_3'y_4',$$

which is $-y_0'(y_2'^3 + y_3'^3 + y_4'^3 - 3y_2'y_3'y_4') - 3(y_0' - y_1')y_2'y_3'y_4',$

is equal to

$$-27(y_0 - y_1)y_2 y_3 y_4 - 9y_0(y_2^3 + y_3^3 + y_4^3 - 3y_2 y_3 y_4),$$

or
$$-9y_0(y_2^3 + y_3^3 + y_4^3) + 27y_1 y_2 y_3 y_4.$$

The transformed polynomial whose vanishing gives the equation of the primal is thus nine times the original polynomial.

The transformation C is an even permutation (243), of y_1, y_2, y_3, y_4; the transformation D is the odd permutation (12), of these. It may be shewn (see §(20) below) that all the possible twenty-four substitutions of y_1, y_2, y_3, y_4 among themselves are obtainable by combinations of the two substitutions (243), (12). But the combinations of C, D, S (for the notation cf. pp. 75, 76),

$$L = DS^2 D, \quad M = S^2, \quad N = CB^2 S^2 C B^2,$$

we at once see change (y_1, y_2, y_3, y_4) into

$$(y_1, \epsilon y_2, \epsilon y_3, \epsilon y_4), \quad (\epsilon y_1, y_2, \epsilon y_3, \epsilon y_4), \quad (\epsilon y_1, \epsilon y_2, y_3, \epsilon y_4)$$

respectively; thus, if α, β, γ, δ be each 0, 1, or 2, subject to $\alpha + \beta + \gamma + \delta \equiv 0 \pmod{3}$, the transformation $L^{\delta - \alpha} M^{\delta - \beta} N^{\delta - \gamma}$ changes $(y_1, \ldots, y_4)$ into $(\epsilon^\alpha y_1, \epsilon^\beta y_2, \epsilon^\gamma y_3, \epsilon^\delta y_4)$. Combining such transformations with those of the symmetric group of twenty-four substitutions obtainable from C and D, as we have seen, we see that, by combination of C, D, S, $(y_0, y_1, \ldots, y_4)$ can be changed into any one of the 24.27 sets given by

$$(y_0, \epsilon^\alpha y_l, \epsilon^\beta y_m, \epsilon^\gamma y_n, \epsilon^\delta y_p),$$

where l, m, n, p is any permutation of 1, 2, 3, 4. In passing we may remark that $D = p(A_0 B)$, and $CB^2 = p(56)$. (Cf. p. 79.)

We have shewn, however (§ (13)), that by combination of the transformation B with the transformations which we now see to arise from C, D and S, we can obtain forty equations of the primal, all of the same form, that is, forty sets of values of $y_0, y_1, \ldots, y_4$ satisfying the equation, in each of which $y_0 = 0$ is one of the Steiner solids, and $y_1 = 0, \ldots, y_4 = 0$ are associated primes, defining a Steiner tetrahedron. There is thus a group of 24.27.40, or $2^3 . 3^4 . 40$ linear transformations of the primal into itself, obtainable by combination of B, C, D, S. Any such transformation must clearly interchange the forty-five nodes among themselves; and it may be shewn by individual examination that the permutation of the nodes due to any one of B, C, D, S is equivalent to an *even* number of transpositions of the nodes. The $2^3 . 3^4 . 40$ transformations thus form a group of even substitutions of the nodes among themselves. Conversely it appears clear, without appeal to the theory of linear transformations of the periods of the theory functions on which Burkhardt relies, that the $2^3 . 3^4 . 40$ transformations exhaust the possible linear self-transformations of the equation

$$y_0^4 - y_0(y_1^3 + \ldots + y_4^3) + 3y_1 y_2 y_3 y_4 = 0.$$

For the transformations of this which leave y_0 unaltered must leave both $y_1^3 + \ldots + y_4^3$ and $y_1 y_2 y_3 y_4$ unaltered, and so be coextensive with the 24.27 transformations of y_1, y_2, y_3, y_4 into forms $\epsilon^\alpha y_l$, $\epsilon^\beta y_m$, $\epsilon^\gamma y_n$, $\epsilon^\delta y_p$ which we have found by combination of C, D, S alone. Also, this equation shews that any possible form of y_0 must be such that $y_0 = 0$ is a Steiner solid, meeting the primal in four planes. We have shewn (§(13)) that it is possible, by combination of B, C, D, S, to obtain forty such Steiner solids $y_0 = 0$. It is only necessary then to assume that there are no other Steiner solids than those we have specified. To prove this is an algebraic problem of which we give no formal solution.

We have also found (§(7)) particular linear transformations of the primal into itself of which each is a projection from a node. Each of these leaves this node, and twelve other nodes unaltered, but interchanges the pairs of nodes lying on the sixteen κ-lines that pass through the centre of projection. Such projection thus gives an even substitution of the nodes and leaves the equation of the primal unaltered. It is to be expected then that every one of the forty-five projections is expressible as a combination of B, C, D, S. We give no formal proof that this is so, though we obtain several examples of this; such proof is probably to be constructed by means of the relations, given in §(7), among projections from nodes lying on a κ-line. But we give below (§(19)) expressions of all of B, C, D, S in terms of projections, in forms necessarily not unique. Presumably then, all the forty-five projections are obtainable by combinations of the four expressions so found. We give some details below (§(19)) in regard to the particular case arising when we consider only the eighteen projections from the nodes lying in a Steiner solid.

(18) **Five subgroups of the group of $2^3.3^4.40$ transformations.** It appears from what we have said that for each of the forty perfectly similar Steiner solids there is a subgroup of the complete group, of order $2^3.3^4$, leaving this solid unaltered; for the solid $y_0 = 0$, this is obtainable by combination of the transformations C, D, S. The transformations of this subgroup will correspond to permutations of the eighteen nodes in this solid.

But the transformations of the subgroup (C, D, S), though effecting even interchanges of the forty-five nodes, do not give solely even interchanges of the eighteen nodes of the solid among themselves. We find on examination that the transformations C and S do so, but the transformation D causes seven transpositions of nodes of the solid, and nine transpositions of nodes not belonging to this solid. For D is $p(A_0 B)$, and there are seven κ-lines from the node $(A_0 B)$ containing nodes of the solid, and nine κ-lines containing nodes not belonging thereto.

The $2^3 . 3^4 . 40$ transformations of the complete group can be supposed, in the familiar way, to consist of the transformations of this subgroup (C, D, S), taken respectively with thirty-nine other transformations, each such transformation corresponding to one of the thirty-nine other Steiner solids. The existence of this subgroup (C, D, S), of *index* forty, arises from the fact that the Burkhardt configuration contains forty subsidiary configurations, all exactly similar, namely the Steiner tetrahedra. The complete group arises by the combination of the transformations of one such configuration into itself, with the transformations which interchange the subsidiary configurations among themselves. Every such set of mutually similar subsidiary configurations likewise gives rise to a subgroup, consisting of the transformations which leave one of these subsidiary configurations unaltered, with index equal to the number of such configurations.

The subsidiary configurations which naturally arise for consideration after the Steiner solids are the Jacobian planes, also forty in number. We expect then to find a subgroup of the complete group leaving any Jacobian plane unaltered and transforming the nine nodes of this plane among themselves, the number of transformations of all the nodes so arising being $2^3 . 3^4 . 40/40$ or $24 . 27$. We approach this result in two ways; first by considering briefly how the general formulae we have given affect a particular Jacobian plane, which we take to be $y_0 = 0 = y_1$; and then, geometrically, by considering the permutations of the nodes of this plane among themselves. From the general formulae of §(17), we see that the plane $y_0 = 0 = y_1$ is changed into itself by the transformations B, C and S; and that these transform y_2, y_3, y_4 respectively by the scheme

	B	C	S
y_2'	$y_2 + y_3 + y_4$	y_4	y_2
y_3'	$y_2 + \epsilon y_3 + \epsilon^2 y_4$	y_2	$\epsilon^2 y_3$
y_4'	$y_2 + \epsilon^2 y_3 + \epsilon y_4$	y_3	$\epsilon^2 y_4$

The nodes in this plane are easily seen to be given by the values of y_2, y_3, y_4 which satisfy the two equations $y_2^3 + y_3^3 + y_4^3 = 0$, $y_2 y_3 y_4 = 0$. It is to be shewn that the combination of the transformations B, C, S leads to a group of 24.27 transformations of y_2, y_3, y_4, each leading to a substitution of the nine nodes among themselves, or leaving these nodes unchanged. Consider, also, the possible interchanges of the nodes; these are the inflexions of a pencil of cubic curves in the plane (y_2, y_3, y_4). It is easy to see that if three of these inflexions, not lying in a line, say P, Q, R, be given, as well as one of the cubic curves of the pencil, then the other six inflexions can be constructed; for the line QR, by its further intersection with the curve, determines another inflexion, say P'; likewise the lines RP and PQ each determine two other inflexions, say Q' and R'; and then the line PP' determines a further inflexion, say P'', and likewise the lines QQ' and RR' each determine another inflexion. Thus all the nine inflexions are determined. There are $9!/3!6!$, or 84 sets of three inflexions possible from the nine, and we know that there are twelve sets of three collinear inflexions. Thus such a triangle as PQR can be chosen in seventy-two ways. We can therefore suppose the set P, Q, R to be made to coincide in turn with seventy-two sets of three inflexions which form a triangle, including itself. And the inflexions forming the angular points of such a triangle can be interchanged among themselves in $3!$ ways. Wherefore, the permutations of the nine inflexions among themselves are 72.6, or 432, in number; of these, one half, or 216, will each be equivalent to an even number of transpositions of two inflexions.

But the equation of the primal is

$$y_0^4 - y_0 y_1^3 - [y_0(y_2^3 + y_3^3 + y_4^3) - 3y_1 y_2 y_3 y_4] = 0,$$

wherein $y_0^4 - y_0 y_1^3$ is unaltered by the transformations C, S, and only multiplied by nine under the transformation B. The transformations B, C, S then, by their operation on y_2, y_3, y_4, change the pencil of cubic curves given by $\lambda(y_2^3 + y_3^3 + y_4^3) + \mu y_2 y_3 y_4 = 0$ among themselves; the changes in the two polynomials

$$y_2^3 + y_3^3 + y_4^3, \quad y_2 y_3 y_4$$

involve certain changes in λ and μ (or y_0 and $-3y_1$), which are those associated, in B, C, S, with the changes in y_2, y_3, y_4. We can compute, however, that the combination $(SC)^3$, while leaving y_0, y_1 unaltered, replaces y_2, y_3, y_4 by ϵy_2, ϵy_3, ϵy_4, so that the repetition of this leads to $\epsilon^2 y_2$, $\epsilon^2 y_3$, $\epsilon^2 y_4$; these changes, though arising in the set of transformations of $y_0, y_1, ..., y_4$, are ineffective as displacements of the nine inflexions. It may be shewn that the transformations B, C, S each lead to an even permutation of the nine nodes. We infer, therefore, as the even displacements of the inflexions are 216 in number, that the transformations B, C, S generate a group of 3.216 transformations. And this is the number, 24.27, which we had reason to expect.

We have seen that the complete group of 24.27.40 transformations is obtainable by combinations of B, C, S and D. It can thus be built up from the subgroup (B, C, S) which we have considered, associated with proper combinations of B, C, S and D. The transformation D, equivalent to the projection $p(A_0 B)$, will leave unaltered the eight Jacobian planes which pass through $(A_0 B)$, while interchanging the remaining thirty-two such planes.

A third subsidiary configuration occurring in the figure is that of one of the forty-five entirely similar Jordan primes. It is to be expected then that the transformations of $y_0, y_1, ..., y_4$ have a subgroup of order $2^3.3^4.40/45$, or 576, interchanging the twelve nodes of a Jordan prime. We examine the case of the prime

$[A_0 B] = 0$. This prime, we easily see, is given by $y_3 - y_4 = 0$, and is unaltered by the transformations B, D, S of § (17). The nodes in this prime are the angular points, other than (AB_0), of the pentahedra $\{A\}$, $\{B_0\}$, $\{AB\}$. If the faces of the three tetrahedra respectively formed by these nodes have equations $\alpha_i = 0$, $\beta_i = 0$, $\gamma_i = 0$, for $i = 1, 2, 3, 4$, there exists, (§ (5)), an identity of the form

$$A\alpha_1\alpha_2\alpha_3\alpha_4 + B\beta_1\beta_2\beta_3\beta_4 + C\gamma_1\gamma_2\gamma_3\gamma_4 = 0,$$

where A, B, C are constants, and in general only one such identity. There is therefore no derangement of the twelve nodes possible among themselves in which this separation into three sets of four nodes is disturbed; there cannot by such derangement be any interchange of the angular points of one tetrahedron with the angular points of another, other than such as arise from interchange of the two tetrahedra. The only possible derangements arise therefore by the interchange of the angular points of each tetrahedron among themselves, coupled with the interchanges of the three tetrahedra among themselves. Now the even interchanges of the angular points of a single tetrahedron among themselves are twelve in number; and there are six possible orders in which the three tetrahedra may be arranged. Any transformation of one tetrahedron involves, under B, D, S, definite derangements of the other two. Thus we can account for possible even substitutions of the nodes, forming the angular points of the three tetrahedra, which are seventy-two in number.

Inspection of the scheme of four transformations of $y_0, y_1, \dots, y_4$ given in § (17) shews that the prime $y_3 - y_4 = 0$ is unaltered by the three transformations B, D, S. But, in fact, these transformations, which are transformations of primes, include transformations which change the sign of the left side of the equation of any plane face of a tetrahedron lying in $y_3 - y_4 = 0$. Thus the combination of B, D, S leads to 72.8 transformations of $y_0, y_1, \dots, y_4$ in all, that is 576, as forecasted, a transformation which changes the sign of every face being counted as identical.

We examine this result now in further detail. The equation of the primal $y_0^4 - y_0(y_1^3 + \ldots) + 3y_1 y_2 y_3 y_4 = 0$ is the same as

$$y_0^4 - y_0\left[y_1^3 + y_2^3 + 2\left(\frac{y_3 + y_4}{2}\right)^3 \right]$$

$$+ 3y_1 y_2\left(\frac{y_3 + y_4}{2}\right)^2 - \frac{3}{4}(y_3 - y_4)^2 [y_0(y_3 + y_4) + y_1 y_2] = 0;$$

we are concerned with the tetrahedra in the prime $[A_0 B] = 0$, which is $y_3 - y_4 = 0$; we put then $y_4 = y_3$ and obtain, in the co-ordinates y_0, y_1, y_2, y_3, the equation

$$y_0^4 - y_0[y_1^3 + y_2^3 + 2y_3^3] + 3y_1 y_2 y_3^2 = 0,$$

representing the quartic surface which is the section of the primal by this prime. The coordinates y_0, y_1, y_2, y_3 of the twelve nodes of this surface, which are the angular points of the three tetrahedra arising from the pentahedra $\{A\}$, $\{B_0\}$, $\{AB\}$, are found to be those given by the rows of the scheme

$\{A\}$				$\{B_0\}$				$\{AB\}$			
0,	1,	$-\epsilon$,	0	0,	1,	$-\epsilon^2$,	0	0,	1,	-1,	0
1,	ϵ^2,	1,	ϵ^2	1,	ϵ,	1,	ϵ	1,	1,	1,	1
1,	ϵ,	ϵ^2,	1	1,	1,	ϵ^2,	ϵ^2	1,	ϵ^2,	ϵ^2,	ϵ
1,	1,	ϵ,	ϵ	1,	ϵ^2,	ϵ,	1	1,	ϵ,	ϵ,	ϵ^2

Further, the transformations B, D, S, and the transformation $(DS^2)^2$, which we denote by E, as they affect y_0, y_1, y_2, y_3, are given by

	B	D	S	E
y_0'	$y_0 - y_1$	y_0	y_0	y_0
y_1'	$-2y_0 - y_1$	y_2	$\epsilon^2 y_1$	ϵy_1
y_2'	$y_2 + 2y_3$	y_1	y_2	ϵy_2
y_3'	$y_2 - y_3$	y_3	$\epsilon^2 y_3$	$\epsilon^2 y_3$

We see at once that, applied to the coordinates, the transformation S leaves unaltered the angular point (AB_0), the pole of the prime under consideration, and, for the *tetrahedra* $\{A\}$, $\{B_0\}$, $\{AB\}$, gives

$$S\{A\} = \{B_0\}, \quad S\{B_0\} = \{AB\}, \quad S\{AB\} = \{A\},$$

the rows of coordinates of the angular points of these tetrahedra, put down in the scheme above, being preserved. Thus S effects an even substitution of the twelve nodes (just as the substitution (abc), for three letters, is equivalent to two transpositions). Further, we find, if (X, Y, Z, T), (X_1, Y_1, Z_1, T_1), (X_2, Y_2, Z_2, T_2) denote the angular points of the three tetrahedra, that the transformation D, while leaving unaltered the angular points of $\{AB\}$, changes (X, Y, Z, T) respectively into (X_1, Z_1, T_1, Y_1), and so changes the tetrahedron $\{A\}$ into the tetrahedron $\{B_0\}$, and conversely, since $D^2 = 1$. Thus S, D together effect the six possible permutations of the three tetrahedra among themselves.

We can then limit our consideration to the effect of B, D, S upon one of these, say $\{AB\}$. We find that, acting upon the angular points of this, $B(X_2, Y_2, Z_2, T_2) = (Y_2, X_2, T_2, Z_2)$, so that B is equivalent to the even substitution $(X_2 Y_2)(Z_2 T_2)$ for these angular points; also we find that E, or $(DS^2)^2$, changes X_2, Y_2, Z_2, T_2 respectively into X_2, T_2, Y_2, Z_2, namely, is equivalent to the even substitution $(Y_2 T_2 Z_2)$. Thus B and E, together, effect the twelve even interchanges of X_2, Y_2, Z_2, T_2 (see below § (20)). But the planes of the tetrahedron $\{AB\}$ opposite to X_2, Y_2, Z_2, T_2 are found to be given respectively by $\mu_0 = 0$, $\mu_1 = 0$, $\mu_2 = 0$, $\mu_3 = 0$, where

$$\mu_0 = y_1 - y_2, \qquad\qquad \mu_1 = 2y_0 + y_1 + y_2 + 2y_3,$$

$$\mu_2 = 2y_0 + \epsilon(y_1 + y_2) + 2\epsilon^2 y_3, \quad \mu_3 = 2y_0 + \epsilon^2(y_1 + y_2) + 2\epsilon y_3;$$

by applying the substitution D the values of μ_1, μ_2, μ_3 are left unaltered, but the sign of μ_0 is changed. Choose then, from the twelve even substitutions of $\mu_0, \mu_1, \mu_2, \mu_3$, that one, say D_i, which interchanges μ_0 and μ_i and at the same time interchanges the other two; then the substitution $D_i D D_i$ changes the sign of μ_i, leaving all others unaltered. Hence, by combining B, D, S, we can change $(\mu_0, \mu_1, \mu_2, \mu_3)$ into any one of the sixteen

$$(\pm\mu_0, \pm\mu_1, \pm\mu_2, \pm\mu_3),$$

equivalent, for our purpose, to eight cases, the $\mu_0, ..., \mu_3$ being taken in any one of the twelve orders obtainable by even substitutions. In this way, it appears that the subgroup (B, D, S), acting on the coordinates $y_0, y_1, ..., y_4$, contains $6.12.8$, or 576 transformations, as was stated. We may add a table of the precise effects of B, E, D, acting on the faces of the tetrahedron $\{AB\}$:

	B	E	D
μ_0'	$-\mu_1$	$\epsilon\mu_0$	$-\mu_0$
μ_1'	$-3\mu_0$	μ_2	μ_1
μ_2'	$(1-\epsilon)\mu_3$	μ_3	μ_2
μ_3'	$(1-\epsilon^2)\mu_2$	μ_1	μ_3

We have examined the effect of the subgroup (B, D, S) upon the nodes in the prime $[A_0 B] = 0$, opposite to (AB_0). The fourth of the original transformations C can be shewn to be equivalent to $p(56)p(AB_0)$, that is, to projection from the node (AB_0), followed by projection from the node (56). The projection $p(AB_0)$ leaves all the nodes in $[A_0 B] = 0$ unaltered. By combination of C with transformations from the subgroup (B, D, S), all the 45.576 transformations of the complete group are to be obtained.

For a fourth subgroup, we next examine the entirely similar configurations which we have called Jordan pentahedra. As these are twenty-seven in number, we expect that the equations for the transformation of the primes $y_0 = 0, ..., y_4 = 0$, in the general group, will lead to a subgroup of index 27, that is, of order $2^3 . 3^4 . 40 \div 3^3$, or 24.40, effecting this number of substitutions among such of the forty-five nodes as may vary when we postulate that a particular pentahedron shall remain unaltered.

Consider in particular the pentahedron $\{AB\}$, of which we denote the prime faces by $\sigma_0 = 0$, $\sigma_1 = 0$, ..., $\sigma_4 = 0$, taking $\sigma_0 = [AB_0]$. There are sixty substitutions of these faces. The transformation D, which is equivalent to projection from the node $(A_0 B)$, opposite to $[AB_0] = 0$, changes the sign of $[AB_0]$, leaving $\sigma_1, ..., \sigma_4$ unaltered; and if D_i be a substitution, from among the sixty, of the form $(\sigma_0 \sigma_i)(\sigma_j \sigma_k)$, which effects a transposition of σ_0 and σ_i, the substitution $D_i D D_i$ changes the sign of μ_i, but leaves the other four primes unaltered. Thus, by combination of D with the sixty even substitutions of the prime faces, we obtain from $\sigma_0, \sigma_1, ..., \sigma_4$ all the values $\sigma_0, \pm \sigma_1, ..., \pm \sigma_4$. By such combination we thus obtain 60.16 even transformations, namely the 24.40 which we expected.

We consider this now in more detail: For the faces of $\{AB\}$ put

$$\sigma_0 = [AB_0], \qquad \sigma_1 = [A_0 B], \qquad \sigma_2 = (\epsilon^2 - \epsilon)[14],$$

$$\sigma_3 = (\epsilon^2 - \epsilon)[25], \quad \sigma_4 = (\epsilon^2 - \epsilon)[36],$$

which, by the formulae connecting $x_1, ..., x_6$ and $y_0, ..., y_4$ (§ (13)), are $\sigma_0 = y_1 - y_2$, $\sigma_1 = -y_3 + y_4$, together with

$$\sigma_2 = \frac{1}{i\sqrt{(3)}} [2y_0 + y_1 + y_2 + y_3 + y_4],$$

$$\sigma_3 = \frac{1}{i\sqrt{(3)}} [2y_0 + \epsilon(y_1 + y_2) + \epsilon^2(y_3 + y_4)],$$

$$\sigma_4 = \frac{1}{i\sqrt{(3)}} [2y_0 + \epsilon^2(y_1 + y_2) + \epsilon(y_3 + y_4)];$$

use also $E = (DS^2)^2$, $F = (DC^2)^2$. Then, under B, D, E, F, we find the transformations:

	B	D	E	F
σ_0'	$-\sigma_2$	$-\sigma_0$	$\epsilon\sigma_0$	σ_1
σ_1'	σ_1	σ_1	$\epsilon^2\sigma_1$	σ_0
σ_2'	σ_0	σ_2	σ_3	σ_2
σ_3'	$\epsilon^2\sigma_4$	σ_3	σ_4	σ_4
σ_4'	$-\epsilon\sigma_3$	σ_4	σ_2	σ_3

Apart then from roots of unity, the transformations B, E, F are equivalent respectively to substitutions for $\sigma_0, \sigma_1, ..., \sigma_4$ which we may denote by $(\sigma_0 \sigma_2)(\sigma_3 \sigma_4)$, $(\sigma_2 \sigma_3 \sigma_4)$, $(\sigma_0 \sigma_1)(\sigma_3 \sigma_4)$; and it can be shewn (§ (20)) that these together generate the sixty even substitutions of $\sigma_0, \sigma_1, ..., \sigma_4$. Thus, as shewn above, by combination of B, E, F, D, we obtain a subgroup of 60.16 even substitutions leaving the pentahedron $\{AB\}$ unaltered. From the

geometrical considerations, this constitutes the aggregate from
the 60.16.27 transformations of the complete group, which has
this property. It is thus possible to find, from the complete group,
twenty-seven transformations, including identity, each leading
to a change of pentahedron, and the combination of these with
the transformations of the subgroup, leads to all the transformations of the complete group. One such of these twenty-seven
transformations is that denoted in § (16) by $(x') = \psi(x)$, given
by the combination of two projections $p(AB_0)\,p(B_0\,C)$, which
transforms the pentahedron $\{A\,B\}$ into $\{A\}$ (and conversely,
since $\psi^2 = 1$). We shall, with reference to this, remark below
(§ (21)), on the self transformations of the pentahedron $\{A\}$
corresponding to those here considered for $\{A\,B\}$.

A fifth subgroup of our general group is that which leaves unaltered one of the thirty-six double sixes of Jordan pentahedra
considered above. Such a subgroup will be of order $2^3 . 3^4 . 40 \div 36$,
or 720; that is 6 !. In particular, the double six whose two rows
consist of the pentahedra $\{A\}, \dots, \{F\}$; $\{A_0\}, \dots, \{F_0\}$, is symmetrical in $x_1, \dots, x_6$, and is unchanged by any of the 720 permutations of these. And clearly this symmetric group leaves
unaltered the equation of the primal expressed in $x_1, x_2, \dots, x_6$. But
it is to be remarked, and will be made clear in § (20) below, that a
single transposition of two of $x_1, \dots, x_6$, besides possibly altering
the order of the columns in the double six, necessarily effects a
transposition of the two rows. Such a transposition, though of
odd character regarded as a substitution of $x_1, \dots, x_6$, is an even
substitution of the nodes of the primal, being (§ (7)) equivalent to
a projection from a node. If we assume that any even substitution of the nodes is obtainable by a combination of the fundamental transformations B, C, D, S (and, for this, we have given a
geometrical justification in § (17)), it follows that the symmetrical
group of 720 substitutions of $x_1, \dots, x_6$ is so obtainable; thus, as
we have seen, it forms a subgroup of the complete group of
$2^3 . 3^4 . 40$ transformations. Individual examples of the expression
of transpositions of $x_1, \dots, x_6$ in terms of B, C, D, S are given by

$p(14) = B^{-1}DB$, $p(56) = CB^2$; the former we have proved (§ (13)); the latter arises by combining the two results

$$C = p(56)\,p(AB_0), \quad B^2 = p(AB_0),$$

which can be verified.

However, another group of 720 operations which leaves a double six unaltered is that of the substitutions of the six columns among themselves (accompanied possibly by transpositions of the pentahedra in one or more columns). The consideration of the relation of this group with that of the symmetrical group of substitutions of $x_1, \ldots, x_6$ arises below (§ (20)).

The other thirty-five double sixes are obtainable from the one considered here by the formulae given in § (16).

(19) **The expression of the fundamental transformations B, C, D, S as transformations of $x_1, \ldots, x_6$. The expression of B, C, D, S in terms of nodal projections.** By the formulae of § (13), the transformations B, C, D, S, so far expressed as transformations for $y_0, y_1, \ldots, y_4$, can be expressed as transformations for $x_1, \ldots, x_6$. Neglecting a common constant factor affecting the transformed values $x_1', \ldots, x_6'$, we thus find:

For B,
$$
\begin{cases}
x_1' = x_1 + \epsilon x_3 + \epsilon x_6, & -x_3' = x_2 + \epsilon^2 x_4 + \epsilon^2 x_5, \\
x_5' = \epsilon x_1 + x_3 + \epsilon x_6, & -x_4' = \epsilon^2 x_2 + x_4 + \epsilon^2 x_5, \\
x_2' = \epsilon x_1 + \epsilon x_3 + x_6, & -x_6' = \epsilon^2 x_2 + \epsilon^2 x_4 + x_5,
\end{cases}
$$

which, save for preceding interchanges among $x_1, \ldots, x_6$ and subsequent interchanges among $x_1', \ldots, x_6'$, is the transformation $(x') = \chi(x)$ used in § (3).

For C,
$$
\begin{cases}
x_1' = x_1 - \tfrac{1}{3}[A_0 B], & x' = x_4 - \tfrac{1}{3}[A_0 B], \\[2mm]
x_2' = x_2 - \dfrac{\epsilon^2}{3}[A_0 B], & x_5' = x_6 - \dfrac{\epsilon}{3}[A_0 B], \\[2mm]
x_3' = x_3 - \dfrac{\epsilon}{3}[A_0 B], & x_6' = x_5 - \dfrac{\epsilon^2}{3}[A_0 B],
\end{cases}
$$

and, since (AB_0) is $(1, \epsilon^2, \epsilon, 1, \epsilon^2, \epsilon)$, by §(1), this represents projection from (AB_0), followed by transposition of x_5' and x_6' (see §(7)); so that we may write $C = p(56)\,p(AB_0)$.

For D,

$$
\begin{cases}
x_1' = x_1 - \tfrac{1}{3}[AB_0], & x_2' = x_2 - \dfrac{\epsilon}{3}[AB_0], & x_3' = x_3 - \dfrac{\epsilon^2}{3}[AB_0], \\[2ex]
x_4' = x_4 - \tfrac{1}{3}[AB_0], & x_5' = x_5 - \dfrac{\epsilon}{3}[AB_0], & x_6' = x_6 - \dfrac{\epsilon^2}{3}[AB_0],
\end{cases}
$$

which shews that $D = p(A_0 B)$.

For S,
$$
\begin{cases}
x_1' = x_1 + \epsilon x_2 + \epsilon x_3, & -x_5' = x_4 + \epsilon^2 x_5 + \epsilon^2 x_6, \\
x_2' = \epsilon x_1 + x_2 + \epsilon x_3, & -x_6' = \epsilon^2 x_4 + x_5 + \epsilon^2 x_6, \\
x_3' = \epsilon x_1 + \epsilon x_2 + x_3, & -x_4' = \epsilon^2 x_4 + \epsilon^2 x_5 + x_6,
\end{cases}
$$

which is clearly $(x') = \chi(x)$, followed by change of x_4', x_5', x_6' respectively into x_5', x_6', x_4'. Thus we may write

$$S = p(45)\,p(46)\,\chi.$$

In replacing a transformation by a product of two (or more) others, some care is necessary in regard to notation. Let U, V denote two transformations, say of variables $z_1, z_2, \dots$; let V be such as to change $z_1, z_2, \dots$ respectively to certain functions of these, which we may denote by $z_1', z_2', \dots$, a fact we denote by $V(z_1, z_2, \dots) = (z_1', z_2', \dots)$; let U be such as to change $z_1', z_2', \dots$ respectively to certain functions of these, which we may denote by $z_1'', z_2'', \dots$, a fact we denote by $U(z_1', z_2', \dots) = (z_1'', z_2'', \dots)$. Then we denote by $UV(z_1, z_2, \dots)$ the value of $U\{V(z_1, z_2, \dots)\}$ regarded as $U(z_1', z_2', \dots)$, which is $(z_1'', z_2'', \dots)$. In effect, the product UV, acting on z_i, is defined as $U(z_i')$.

But, if $f(z_1, z_2, \dots)$ denote a function of $z_1, z_2, \dots$, it is a usual convention to denote $f(z_1', z_2', \dots)$ by $Vf(z_1, z_2, \dots)$. This convention, because $UV(z_1, z_2, \dots) = (z_1'', z_2'', \dots)$, then leads to

$$UVf(z_1, z_2, \dots) = f(z_1'', z_2'', \dots).$$

Now, if $\zeta_1, \zeta_2, \dots$ be values of $z_1, z_2, \dots$, let the function

$$f[U(\zeta_1), U(\zeta_2), \dots],$$

regarded as a function of $\zeta_1, \zeta_2, \ldots$, be momentarily denoted by $\phi(\zeta_1, \zeta_2, \ldots)$. Then, by the same convention,

$$V\{Uf(z_1, z_2, \ldots)\}$$
$$= V\{f[U(z_1), U(z_2), \ldots]\} = V\phi(z_1, z_2, \ldots) = \phi[V(z_1), V(z_2), \ldots]$$
$$= f[U(V(z_1)), U(V(z_2)), \ldots] = f[U(z_1'), U(z_2'), \ldots] = f(z_1'', z_2'', \ldots)$$
$$= f[UV(z_1), UV(z_2), \ldots] = (UV)f(z_1, z_2, \ldots).$$

Hence, the result of the product transformation UV, acting on $f(z_1, z_2, \ldots)$, is obtained by operating on $f(z_1, z_2, \ldots)$ *first* by U, and then, *regarding what is obtained as a function of* $z_1, z_2, \ldots$, acting on this by V. The applications arising here are generally when $f(z_1, z_2, \ldots)$ is a linear function of its variables. In particular this rule, applied for example to z_i, requires us to operate with V upon $U(z_i)$, *regarded as a function of* z_i, and thus gives $U(z_i')$; which is the result here taken as fundamental.

This being understood, we remark in regard to the transformations B, C, D, S,

(1) B; that

$$B^2 = p(AB_0), \quad B^4 = 1, \text{ so that } B = p(AB_0) . B . p(AB_0).$$

It is also convenient to use a transformation we denote by B_1, which may be defined as $p(B_0 C) . B . p(B_0 C)$; for this then also $B_1^2 = p(AB_0)$, since (§ (7)), the projections $p(AB_0)$, $p(B_0 C)$ are commutable. Various possible expressions for B_1 in terms of projections are then

$$B_1 = p(AD_0)\,p(AF_0)\,p(25)\,p(36)$$
$$= p(AF_0)\,p(FB_0)\,p(EB_0)\,p(AD_0)$$
$$= p(AD_0)\,p(DB_0)\,p(CB_0)\,p(AF_0)$$
$$= p(AE_0)\,p(B_0 D)\,p(A_0 B)\,p(B_0 C)$$
$$= p(AC_0)\,p(A_0 B)\,p(AF_0)\,p(B_0 E)$$
$$= p(AD_0)\,p(DB_0)\,p(AF_0)\,p(36),$$

while

$$B = p(B_0 C)\,p(AE_0)\,p(B_0 D)\,p(A_0 B)$$
$$= p(B_0 D)\,p(A_0 B)\,p(B_0 F)\,p(AC_0) = \text{etc.}$$

For the proof of these we may, by § (13), express the projections by the variables $y_0, y_1, ..., y_4$; but the equivalence of any two of the forms follows by the rules given in § (7).

(2) C; we have

$$C^3 = 1, \quad C = p(56)\,p(AB_0) = p(E_0 F)\,p(56) = p(AB_0)\,p(E_0 F),$$

so that $\qquad\qquad p(56) = CB^2, \quad (CB^2)^2 = 1.$

(3) D; we have already remarked that

$$D = p(A_0 B), \quad D^2 = 1.$$

(4) S; we have $S^3 = 1$; and various forms are given by

$$S = p(BC_0)\,p(DE_0)\,p(EF_0)\,p(CA_0)$$
$$= p(DE_0)\,p(EF_0)\,p(AB_0)\,p(CA_0)$$
$$= p(BC_0)\,p(CA_0)\,p(13)\,p(12) = p(13)\,p(12)\,p(AB_0)\,p(BC_0).$$

We can prove that S is commutable with B_1, or $SB_1 = B_1 S$, and defining a transformation, say μ, by

$$\mu = p(12)\,p(13)\,p(BC_0) = p(CA_0)\,p(12)\,p(13)$$
$$= p(EF_0)\,p(DE_0)\,p(CA_0),$$

we have $\qquad S = \mu^2, \quad \mu B_1 = B_1\mu, \quad \mu^3 = B_1^2 = p(AB_0),$

the expression of μ in the variables $y_0, y_1, ..., y_4$ being

$$(y_0', y_1', ..., y_4') = (y_0, \epsilon y_1, y_2, \epsilon y_4, \epsilon y_3).$$

(5); the transformations E, F are given by

$$E = (DS^2)^2 = p(12)\,p(13)\,p(45)\,p(46), \quad F = (DC^2)^2 = p(23)\,p(56).$$

(6); we have given in § (16) the expression in terms of $x_1, ..., x_6$ of a transformation ψ, which may be defined by

$$\psi = p(AB_0)\,p(B_0 C).$$

In terms of $y_0, y_1, ..., y_4$ its expression is

$$(y_0', ..., y_4') = (y_0, \epsilon y_2, \epsilon^2 y_1, y_4, y_3).$$

This transforms the pentahedra $\{A\}$, $\{AB\}$ into one another, being such that $\psi^2 = 1$.

(7); the transformation χ several times used, ($\S$(3)), expressed in terms of $(y_0, ..., y_4)$ is given by

$$(y_0', ..., y_4') = (y_0, \epsilon^2 y_1, \epsilon y_2, \epsilon y_3, \epsilon^2 y_4),$$

which shews that $\chi^3 = 1$. Also we can prove that

$$\psi = \chi p(16)\, p(24)\, p(35) = \chi p(AC_0)\, p(A_0 C),$$
$$\chi = p(AB_0)\, p(B_0 C)\, p(AC_0)\, p(A_0 C)$$
$$= p(BC_0)\, p(A_0 B)\, p(AB_0)\, p(B_0 C).$$

Further relations may also be given here. We have already referred to

$$p(14) = BDB^{-1} = B^{-1}DB, \quad (CS)^3 = (SC)^3 = DS^2 D$$

and we find that $\qquad p(B_0 C) = S^{-1}DS$.

We also introduce a transformation A, which can be defined by

$$A = p(14)\, p(56)\, p(EF_0), = p(14)\, p(EF_0)\, p(A_0 B),$$
$$= p(56)\, p(EF_0)\, p(14),$$

whose expression, therefore, in terms of $x_1, ..., x_6$, is given by

$$x_4' = x_1 - \tfrac{1}{3}[E_0 F], \quad x_2' = x_2 - \frac{\epsilon}{3}[E_0 F], \quad x_3' = x_3 - \frac{\epsilon^2}{3}[E_0 F],$$

$$x_1' = x_4 - \tfrac{1}{3}[E_0 F], \quad x_6' = x_5 - \frac{\epsilon^2}{3}[E_0 F], \quad x_5' = x_6 - \frac{\epsilon}{3}[E_0 F];$$

in terms of $(y_0, y_1, ..., y_4)$ this is given by

$$y_i' = \eta_i - \tfrac{1}{3}u, \text{ for } i = 0, 1, ..., 4,$$

where

$$u = 2y_0 + y_1 + y_2 + y_3 + y_4, \quad (\eta_0, \eta_1, \eta_2, \eta_3, \eta_4) = (y_0, y_3, y_1, y_2, y_4),$$

so that A is $p(14)$ preceded by the change of y_1, y_2, y_3 respectively into y_3, y_1, y_2 (see §(13)); and it can be shewn that $p(56)\, p(EF_0)$ leads to $(y_0', y_1', ..., y_4') = (y_0, y_3, y_1, y_2, y_4)$. We find that

$$AC = (BD)^2, \quad A^2 = CB^2 D = CDB^2 = p(56)\, p(A_0 B),$$
$$A^3 = BDB^{-1} = p(14), \quad A^6 = 1, \quad A = DCBDB,$$

leading to $\qquad C = A^2 BA^3 B, \quad D = B^{-1}A^3 B,$

so that B, C, D are expressible by the two transformations A, B. By virtue of $(CB^2)^2 = 1$, A and B are connected by an identity, as also follows from $A = DCBDB$. Another form for A^2 is $[p(56)\,p(EF_0)]^2$. See *Postscript* on p. 98.

It is convenient for purposes of verification to have the formulae in terms of $(y_0, y_1, ..., y_4)$ for the projections from the eighteen nodes which exist in the Steiner solid $y_0 = 0$, or $x_1 + x_2 + x_3 = 0$. If we denote the projections

$$p(23),\ p(31),\ p(12);\quad p(BC_0),\ p(CA_0),\ p(AB_0);$$
$$p(EF_0),\ p(FD_0),\ p(DE_0),$$

respectively, by

$$X,\ Y,\ Z;\quad P,\ Q,\ R;\quad U,\ V,\ W,$$

and the projections, from the nodes in the opposite edges of the tetrahedron,

$$p(56),\ p(64),\ p(45);\quad p(B_0C),\ p(C_0A),\ p(A_0B);$$
$$p(E_0F),\ p(F_0D),\ p(D_0E),$$

respectively, by

$$X',\ Y',\ Z';\quad P',\ Q',\ R';\quad U',\ V',\ W',$$

we find

	X	Y	Z	P	Q	R	U	V	W
y'_0	y_0	y_0	y_0	y_0	y_0	y_0	y_0	y_0	y_0
y'_1	y_4	ϵy_4	$\epsilon^2 y_4$	y_1	y_1	y_1	y_3	ϵy_3	$\epsilon^2 y_3$
y'_2	y_2	y_2	y_2	y_2	y_2	y_2	y_2	y_2	y_2
y'_3	y_3	y_3	y_3	ϵy_4	$\epsilon^2 y_4$	y_4	y_1	$\epsilon^2 y_1$	ϵy_1
y'_4	y_1	$\epsilon^2 y_1$	ϵy_1	$\epsilon^2 y_3$	ϵy_3	y_3	y_4	y_4	y_4

	X'	Y'	Z'	P'	Q'	R'	U'	V'	W'
y'_0	y_0	y_0	y_0	y_0	y_0	y_0	y_0	y_0	y_0
y'_1	y_1	y_1	y_1	ϵy_2	$\epsilon^2 y_2$	y_2	y_1	y_1	y_1
y'_2	y_3	ϵy_3	$\epsilon^2 y_3$	$\epsilon^2 y_1$	ϵy_1	y_1	y_4	ϵy_4	$\epsilon^2 y_4$
y'_3	y_2	$\epsilon^2 y_2$	ϵy_2	y_3	y_3	y_3	y_3	y_3	y_3
y'_4	y_4	y_4	y_4	y_4	y_4	y_4	y_2	$\epsilon^2 y_2$	ϵy_2

The table shews that the six products YZ, QR, VW, $Y'Z'$, $Q'R'$, $V'W'$ are all 'diagonal' transformations, that is, the transformed

y_i' is a multiple of y_i, so that any two of these six products are commutable; and in particular that

$$Y'Z' . YZ = YZ . Y'Z' = VW . V'W'; \quad YZ(Y'Z')^{-1} = QR . Q'R';$$
$$W = RZR.$$

Other results arising are $YZ = QRVW$, and

$$S = YZQR, \quad DS^2 = YZ . Y'Z' . P' = P' . YZ . Y'Z',$$
$$(DS^2)^2 = (YZ . Y'Z')^2,$$

D, S being the Burkhardt transformations. Also, a consideration of the κ-lines in the figure shews that all the eighteen projections are expressible in terms of four of them.

We have expressed all of B, C, D, S in terms of nodal projections; and anticipated that any of these forty-five projections, being even linear transformations of the nodes, must be conversely expressible in terms of B, C, D, S. For the formal proof of this we expect that the rules given in §(7), depending on the κ-lines, may be sufficient. As results of these rules, every one of X, Y, Z, in the table above, is commutable with any one of X', Y', Z'; every one of P, Q, R with any one of P', Q', R'; and every one of U, V, W with any one of U', V', W'; for example, that P, or $p(BC_0)$, is commutable with Q', or $p(C_0A)$, follows because the nodes (BC_0), (C_0A) are angular points of the same pentahedron. A consequence of this is the existence of equations such as $(XY')^2 = 1$; for this is $XY' . XY'$, or $XXY'Y'$, or $X^2(Y')^2$, which is 1. From these the equations such as $YZ . Y'Z' = Y'Z' . YZ$ follow at once. Other results are not so obvious; as for example that $p(AC_0) p(AD_0)$ is commutable with $p(B_0E) p(B_0F)$; or that

$$p(B_0C) p(A_0B) . p(AE_0) p(B_0D) . p(A_0B) p(B_0C)$$
$$= p(AF_0) p(B_0E).$$

But, as an example of the application of the rules, we may give a proof that $B^2 = p(AB_0)$, a fact which is immediately proved by using the expressions for B and $p(AB_0)$ in terms of $y_0, y_1, ..., y_4$. Define the transformation B_1, by the first form given above, as

$B_1 = p(AD_0)\,p(AF_0)\,p(25)\,p(36)$. Notice that four κ-lines are given by the sets of three nodes

$$(AD_0),\ (FB_0),\ (25);\quad (AF_0),\ (DB_0),\ (25);$$
$$(AD_0),\ (EB_0),\ (36);\quad (AF_0),\ (CB_0),\ (36).$$

Hence
$$p(AD_0)\,p(25) = p(FB_0)\,p(AD_0);$$
$$p(AD_0)\,p(36) = p(EB_0)\,p(AD_0).$$

Therefore
$$B_1 = p(AF_0)\,p(AD_0)\,p(25)\,p(36) = p(AF_0)\,p(FB_0)\,p(AD_0)\,p(36)$$
$$= p(AF_0)\,p(FB_0)\,p(EB_0)\,p(AD_0).$$

Also
$$p(AF_0)\,p(25) = p(DB_0)\,p(AF_0);$$
$$p(AF_0)\,p(36) = p(CB_0)\,p(AF_0).$$

Wherefore
$$B_1 = p(AD_0)\,p(DB_0)\,p(AF_0)\,p(36)$$
$$= p(AD_0)\,p(DB_0)\,p(CB_0)\,p(AF_0).$$

Multiplying the former value of B_1 into the latter, we obtain

$$B_1^2 = p(AF_0)\,p(FB_0)\,p(EB_0)\,p(AD_0)$$
$$.\,p(AD_0)\,p(DB_0)\,p(CB_0)\,p(AF_0).$$

But we have

$$[p(AD_0)]^2 = 1,\quad p(FB_0)\,p(EB_0)\,p(DB_0)\,p(CB_0) = p(AB_0);$$

thus

$$B_1^2 = p(AF_0)\,p(AB_0)\,p(AF_0) = p(AF_0)\,p(AF_0)\,p(AB_0) = p(AB_0).$$

If we now define B by $B = p(B_0C)\,.\,B_1\,.\,p\,(B_0C)$, we obtain

$$B^2 = p(B_0C)\,.\,B_1^2\,.\,p(B_0C)$$
$$= p(B_0C)\,p(AB_0)\,p(B_0C) = p(B_0C)\,p(B_0C)\,p(AB_0),$$

so that
$$B^2 = p(AB_0).$$

That the form here used as definition of B_1, and the definition given above for B, agree with the original definition of B given above, in terms of $y_0, \ldots, y_4$, may be verified by expressing the component projections in terms of these variables; or, by the longer but interesting process of combining the forms for the projections given in § (7) in terms of $x_1, \ldots, x_6$.

(20) **The application of the substitutions of $x_1, \ldots, x_6$ to the twelve pentahedra $\{A\}$, $\{B\}$, ... $\{F_0\}$.** We first state two elementary lemmas.

Lemma I. If u, w be operations respectively of period 3 and 2, so that $u^3 = 1$, $w^2 = 1$, and if also $(uw)^3 = 1$, $(wu)^3 = w(uw)^3 w^{-1} = 1$, then a group of twelve operations is given by

$$1, \quad w, \quad u^{-1}wu, \quad uwu^{-1},$$
$$u, \quad uw, \quad wu, \quad wuw,$$
$$u^{-1}, \quad (uw)^{-1}, \quad (wu)^{-1}, \quad (wuw)^{-1},$$

where, in a product of two operations $\vartheta\phi$, it is meant that ϕ is carried out before ϑ. In this group, the operations in the first row are mutually commutable, all of period 2, forming themselves a group, which is self-conjugate in the group of twelve operations. Those in the second and third rows are all of period 3.

We may call this group the *tetrahedral* group. In particular, considering four numbers 1, 2, 3, 4, if u, w be respectively the substitutions $u = (234)$, $w = (12)(34)$, the group consists of

$$1, \quad (12)(34), \quad (14)(23), \quad (13)(24),$$
$$(234), \quad (132), \quad (124), \quad (143),$$
$$(243), \quad (123), \quad (142), \quad (134).$$

As $(234)(13)(234)(13) = (12)(34)$, we see that the symmetrical group of twenty-four substitutions of four numbers can be generated by combining the two substitutions (13), (234).

Lemma II. The sixty *even* substitutions of five numbers, 1, 2, 3, 4, 5, consist of identity, with twenty substitutions which leave two of the numbers unchanged, and fifteen substitutions which leave one number unchanged, together with twenty-four substitutions consisting of the powers of six cyclical substitutions each involving all the five numbers. The powers of any one of these cyclical substitutions, taken with identity, form, of course, a group of five operations. The six groups thus arising from the six cyclical substitutions are mutually conjugate in pairs.

The first set of twenty substitutions, spoken of, each leaving two numbers unchanged, clearly consist of the ten substitutions

(123), (124), (125), (134), (135), (145), (234), (235), (245), (345),

and the ten squares of these, (132), (142), etc.

The second set, of fifteen substitutions, each leaving one number unaltered, are, also obviously,

$(12)(34)$, $(12)(35)$, $(12)(45)$; $\quad$ $(13)(24)$, $(13)(25)$, $(13)(45)$;

$(14)(23)$, $(14)(25)$, $(14)(35)$; $\quad$ $(15)(23)$, $(15)(24)$, $(15)(34)$;

$$(23)(45), \quad (24)(35), \quad (25)(34).$$

Put now $\qquad \vartheta = (12345)$, $\quad \phi = (12354)$,

which give

$$\vartheta^{-1}\phi^2 = (124), \quad \vartheta\phi\vartheta^{-1} = \phi^2\vartheta^2\phi^{-2} = (15234).$$

Then the remaining twenty-four even substitutions of five numbers, in addition to the $1 + 20 + 15$ already enumerated, are given by

$$\phi^n, \quad \vartheta^n, \quad \phi\vartheta^n\phi^{-1}, \quad \phi^2\vartheta^n\phi^{-2}, \quad \phi^3\vartheta^n\phi^{-3}, \quad \phi^4\vartheta^n\phi^{-4},$$
$$\ldots(n = 1, 2, 3, 4).$$

If we put

$$u = (345), \quad v = (12)(45), \quad w = (23)(45),$$

so that $\quad u^3 = 1$, $\quad v^2 = 1$, $\quad w^2 = 1$, $\quad (uv)^2 = (vu)^2 = 1$,

$$(uw)^3 = (wu)^3 = (vw)^3 = (wv)^3 = 1,$$

then we have $\qquad \vartheta = vwu$, $\quad \phi = vwu^2$,

and it can be verified that every one of the sixty substitutions under consideration can be expressed as a product of powers of the three substitutions u, v and w.

Now, considering the synthemes arising in the twelve systems or *families* $\{A\}$, $\{B\}$, ..., $\{F_0\}$, put down in § (1), we regard any such syntheme, say $(14.36.25)$, as being unaltered by a cyclical change of the three duads which occur therein, so that it is the same as $(36.25.14)$, or $(25.14.36)$. We may make any one of the 720 substitutions of the numbers 1, 2, 3, ..., 6 in the synthemes of any of the twelve families. Such a substitution *will either inter-change the five synthemes which make up the family among them-selves, or it will change these five synthemes into the synthemes of another family, taken in some order.* In particular, therefore, a substitution which leaves a particular syntheme unaltered will either leave unaltered both the families to which this syntheme belongs, save for possible changes in the order of the synthemes, or will interchange these two families.

This is a property of the scheme given in § (1). But we may prove it by recurring to the geometrical results we have de-veloped. For any substitution among $1, 2, ..., 6$ may be regarded as the product of transpositions of twos of the numbers. Consider, for example, the transposition (14). The duad 14 occurs in one of the five synthemes of every one of the twelve families $\{A\}, \{B\}, ..., \{F_0\}$; for instance, in the family $\{A\}$ we have

$$(AB_0) = (14.36.25),$$

and this syntheme occurs also in $\{B_0\}$. In our geometrical point of view, the common angular point (AB_0), of the two pentahedra $\{A\}, \{B_0\}$, is equally an angular point of the pentahedron $\{AB\}$; and the twelve remaining angular points of these three penta-hedra, which lie in the prime $[A_0 B] = 0$, form three tetrahedra of which any two are in perspective with one another from any angular point of the third. In particular, the angular points of the pentahedra $\{A\}, \{B_0\}$ are projections of one another from the node (14), which is an angular point of $\{AB\}$. But, we have seen, § (7), that projection from this node interchanges the two points of the primal which are given by

$$(x_1, x_2, x_3, x_4, x_5, x_6) \text{ and } (x_4, x_2, x_3, x_1, x_5, x_6).$$

The transposition (14) thus interchanges the synthemes of the families $\{A\}, \{B_0\}$; as is obvious also at once by inspection of the scheme in § (1). Equally this transposition interchanges the families $\{A_0\}$ and $\{B\}$. Now, $\{L\}, \{M\}$ denoting any two of the twelve families $\{A\}, ..., \{F_0\}$, use the symbol $\begin{pmatrix} M \\ L \end{pmatrix}$ to denote a substitution among $1, 2, ..., 6$ which changes the synthemes of

$\{L\}$ into the synthemes of $\{M\}$, taken in some order. Then, by what we have said, we may put

$$(14) = \begin{pmatrix} B_0 \\ A \end{pmatrix}, \text{ or } \begin{pmatrix} A \\ B_0 \end{pmatrix}, \text{ or } \begin{pmatrix} A_0 \\ B \end{pmatrix}, \text{ or } \begin{pmatrix} B \\ A_0 \end{pmatrix}.$$

Since the node (14) belongs also to the pentahedra $\{CD\}$ and $\{EF\}$, the transposition (14) can similarly be expressed in terms of C and D_0, or C_0 and D; or in terms of E and F_0, or E_0 and F. A similar discussion may be made for every transposition (ij).

It follows from this that every one of the 360 *odd* substitutions of $1, 2, \ldots, 6$, that is any substitution which is the product of an odd number of transpositions of two of the numbers, changes any one of the six families $\{A\}, \ldots, \{F\}$ into one of the families $\{A_0\}, \ldots, \{F_0\}$, and conversely. Also, that any *even* substitution interchanges the six families $\{A\}, \ldots, \{F\}$ among themselves, and interchanges also the families $\{A_0\}, \ldots, \{F_0\}$ among themselves.

Consider now, first, even substitutions of $1, 2, \ldots, 6$ which leave a particular family, say $\{A\}$, unaltered, save for possible changes in the order of its five synthemes. It is easily seen that there are sixty such substitutions, forming a group isomorphic with the even substitutions of five objects. For, consider the particular syntheme (14.36.25), belonging to $\{A\}$; this syntheme is evidently changed into itself by the substitution $w = (25)(36)$, and is changed into (25.14.36), which we consider equivalent to the original, by the substitution $u = (123)(456)$. These definitions lead to $uw = (126)(345)$, and $wu = (153)(426)$, which are of period 3. Thus, by Lemma I above, the syntheme (14.36.25) is changed into itself by the twelve substitutions of a tetrahedral group; as is obvious also by inspection. Whence, as these twelve are even substitutions, the family $\{A\}$ is changed into itself by these substitutions, by what we have proved; for instance, considering $w = (25)(36)$, the transposition (36) changes $\{A\}$ into $\{B_0\}$, and (25) changes $\{B_0\}$ into $\{A\}$. Similarly, the family $\{A\}$ is changed into itself by the tetrahedral groups arising in the same way from all the five synthemes of $\{A\}$. On examination, we find that the five tetrahedral groups have substitutions in common, and that, including identity, only thirty-six substitutions leaving $\{A\}$ unaltered are obtainable in this way. There are, however, besides these, six cyclical substitutions, of period 5, leaving $\{A\}$ unaltered, each obtained by keeping one of the numbers $1, 2, \ldots, 6$ unaltered, and changing the others cyclically. These give, besides identity, $6 . 4$ or 24 new substitutions leaving $\{A\}$ unaltered. For instance (26435) is such a substitution, as we easily verify. And this fact can also be seen by expressing (26435) as (26) (64) (43) (35), which, by what we have seen, is equivalent to

$$\binom{A}{D_0}\binom{D_0}{B}\binom{B}{C_0}\binom{C_0}{A},$$

and so changes $\{A\}$ into $\{A\}$. The form of the cycle can in fact be read off by considering the duads which occupy the third place in the synthemes of $\{A\}$. The powers of this substitution

$$(26435)^2 = (24563), \quad (26435)^3 = (23654)$$

are similarly obtainable, in fact, by considering the duads which occupy the second place in the synthemes of $\{A\}$; and $(26435)^4$, or $(26435)^{-1}$, can be read off by inspection of the duads in the third place. The cycle which omits the number 2 can similarly be read off by putting the duads in the various synthemes of $\{A\}$ in forms in which 2 occurs in all the first duads, and then considering the second (or third) duads in cyclical order. Similarly for the other four cyclical substitutions. The twenty-four substitutions so obtained from the various cycles, taken with the thirty-six previously spoken of, constitute in fact the group of sixty even substitutions leaving $\{A\}$ unaltered.

We prove this in a more systematic way by considering all the possible 360 even substitutions of $1, 2, \ldots, 6$, shewing that the sixty substitutions we have found are a subgroup of this alternate group, isomorphic with the group of even substitutions of five objects.

The even substitutions of $1, 2, \ldots, 6$ consist, in fact, besides identity, of forty-five substitutions such as (12) (34), together with forty substitutions such as (123) (456), and of 6.24 or 144 cyclical substitutions such as (23456), each keeping one of $1, 2, \dots, 6$ unaltered, to which must be added forty substitutions such as (123), each keeping three of $1, 2, \dots, 6$ unaltered, and also ninety substitutions such as (12)(3456). Now put, as particular even substitutions of $1, 2, \dots, 6$,

$$u = (123)(456), \qquad v = (13)(45), \quad w = (25)(36),$$

with $\quad \vartheta = vwu = (14265), \quad \phi = vwu^2 = (16234),$

and consider the substitutions of the families $\{A\}, \dots, \{F\}$ which arise by applying these particular substitutions. We have

$$u = (12)(23)(45)(56),$$

and the nodes represented by (12), (23), (45), (56) are respectively angular points of the pentahedra $\{AE\}, \{BE\}, \{BE\}, \{AE\}$; in fact 12 is a duad in the synthemes (AE_0), and 23 is a duad in (E_0B), and 45 is a duad in (BE_0), and 56 is a duad in (E_0A), in the scheme of §(1). Thus, by what we have said, the substitution u may be written as

$$u = \begin{pmatrix} A \\ E_0 \end{pmatrix} \begin{pmatrix} E_0 \\ B \end{pmatrix} \begin{pmatrix} B \\ E_0 \end{pmatrix} \begin{pmatrix} E_0 \\ A \end{pmatrix},$$

and leaves the family $\{A\}$ unaltered. We may similarly put u into other forms proper to shew the effect of u upon $\{B\}, \dots, \{F\}$. In particular, (56) is a node of $\{BF\}$ and $\{CD\}$, as inspection of the table in §(1) verifies, and so for the other transpositions. Thus we find the five alternative forms for u,

$$\begin{pmatrix} B \\ D_0 \end{pmatrix} \begin{pmatrix} D_0 \\ C \end{pmatrix} \begin{pmatrix} C \\ F_0 \end{pmatrix} \begin{pmatrix} F_0 \\ B \end{pmatrix}, \quad \begin{pmatrix} C \\ F_0 \end{pmatrix} \begin{pmatrix} F_0 \\ A \end{pmatrix} \begin{pmatrix} A \\ D_0 \end{pmatrix} \begin{pmatrix} D_0 \\ C \end{pmatrix}, \quad \begin{pmatrix} E \\ A_0 \end{pmatrix} \begin{pmatrix} A_0 \\ F \end{pmatrix} \begin{pmatrix} F \\ C_0 \end{pmatrix} \begin{pmatrix} C_0 \\ D \end{pmatrix},$$

$$\begin{pmatrix} F \\ C_0 \end{pmatrix} \begin{pmatrix} C_0 \\ D \end{pmatrix} \begin{pmatrix} D \\ A_0 \end{pmatrix} \begin{pmatrix} A_0 \\ E \end{pmatrix}, \quad \begin{pmatrix} D \\ B_0 \end{pmatrix} \begin{pmatrix} B_0 \\ E \end{pmatrix} \begin{pmatrix} E \\ B_0 \end{pmatrix} \begin{pmatrix} B_0 \\ F \end{pmatrix}.$$

From these six forms of u we see that u leaves each of $\{A\}, \{B\}, \{C\}$ unaltered, but changes $\{D\}$ into $\{E\}$, and $\{E\}$ into $\{F\}$, and $\{F\}$ into $\{D\}$. This result may be expressed by

$$u = (123)(456) = (DEF).$$

Similarly the substitution $v = (13)(45)$, put into forms respectively applicable to the families $\{A\}, ..., \{F\}$, is equal to every one of the six

$$\binom{A}{D_0}\binom{D_0}{A}, \quad \binom{C}{E_0}\binom{E_0}{B}, \quad \binom{B}{F_0}\binom{F_0}{C},$$

$$\binom{D}{A_0}\binom{A_0}{D}, \quad \binom{F}{B_0}\binom{B_0}{E}, \quad \binom{E}{C_0}\binom{C_0}{F},$$

and we may express the result of its operation on $\{A\}, ..., \{F\}$ by writing
$$v = (13)(45) = (BC)(EF).$$

Likewise, $w = (25)(36)$ can be expressed in the six forms

$$\binom{A}{B_0}\binom{B_0}{A}, \quad \binom{B}{A_0}\binom{A_0}{B}, \quad \binom{D}{F_0}\binom{F_0}{C},$$

$$\binom{C}{E_0}\binom{E_0}{D}, \quad \binom{F}{D_0}\binom{D_0}{E}, \quad \binom{E}{C_0}\binom{C_0}{F},$$

so that we can write
$$w = (25)(36) = (CD)(EF).$$

But, in terms of the symbols B, C, D, E, F, these forms for u, v, w are precisely those which we have seen (Lemma II) to generate the alternate group of five letters. These substitutions u, v, w, namely $(123)(456)$, $(13)(45)$, $(25)(36)$ thus generate a group of sixty even substitutions leaving $\{A\}$ unaltered, but altering $\{B\}, \{C\}, \{D\}, \{E\}, \{F\}$ among themselves, by this alternate group. The particular substitutions enumerated above, leaving $\{A\}$ unaltered, are all included in this group.

It is easy to verify that u, v, w, and hence all the combinations of these, equally leave the family $\{A_0\}$ unaltered. The group of sixty substitutions may then be appropriately denoted by (A, A_0).

In this group, respectively derived from $\vartheta = vwu$, $\phi = vwu^2$, are contained the six cyclic substitutions, of order 5, given by

$$\phi = (16234), \qquad \vartheta = (14265), \qquad \phi\vartheta\phi^{-1} = (13256),$$
$$\phi^2\vartheta\phi^{-2} = (26435), \quad \phi^3\vartheta\phi^{-3} = (14532), \quad \phi^4\vartheta\phi^{-4} = (15436),$$

which, excluding identity, together with their powers, provide twenty-four of the sixty substitutions which leave $\{A\}$, or $\{A_0\}$ unaltered; respectively, these leave unaltered the numbers 5, 3, 4, 1, 6, 2. With twenty substitutions of the form $(123)(456)$, and fifteen of the form $(25)(36)$, they make up, with identity, the sixty substitutions of the group (A, A_0).

We can pass from the family $\{A\}$ to the families

$$\{B_0\}, \{C_0\}, \dots, \{F_0\}$$

by making the respective transpositions (14), (16), (13), (12), (15). Each of these is a choice from three possibilities; for instance, as the symbol of the node $(A B_0)$ is $(14.36.25)$, we can pass from $\{A\}$ to $\{B_0\}$ by using either of the transpositions (36), (25) in place of (14). It follows that each of $\{B_0\}, \dots, \{F_0\}$, and, therefore, each of $\{B\}, \dots, \{F\}$, is unaltered by a group of sixty even substitutions; these groups, $(B, B_0), \dots, (F, F_0)$ are made up of substitutions obtainable from those of (A, A_0) by the respective five transpositions (14), (16), $\dots$, (15).

But the six even groups $(A, A_0), \dots, (F, F_0)$, thus arising, contain common substitutions; and the aggregate of their substitutions does not make up the 360 even substitutions of $1, 2, \dots, 6$. For instance, consider the effect of the substitution (123), or $(12)(23)$, upon $\{A\}, \dots, \{F\}$; to obtain these, put the substitution in the forms

$$(12)(23) = \begin{pmatrix} C \\ F_0 \end{pmatrix} \begin{pmatrix} F_0 \\ A \end{pmatrix} = \begin{pmatrix} A \\ E_0 \end{pmatrix} \begin{pmatrix} E_0 \\ B \end{pmatrix} = \begin{pmatrix} B \\ D_0 \end{pmatrix} \begin{pmatrix} D_0 \\ C \end{pmatrix}$$

$$= \begin{pmatrix} F \\ C_0 \end{pmatrix} \begin{pmatrix} C_0 \\ D \end{pmatrix} = \begin{pmatrix} D \\ B_0 \end{pmatrix} \begin{pmatrix} B_0 \\ E \end{pmatrix} = \begin{pmatrix} E \\ A_0 \end{pmatrix} \begin{pmatrix} A_0 \\ F \end{pmatrix},$$

from whence we infer

$$(123) = (ACB)(DFE).$$

Thus this substitution (123), for the numbers $1, 2, \dots, 6$, does not leave any one of the families $\{A\}, \dots, \{F\}$ unaltered; and the same is therefore true for the forty even substitutions of this form. Or again, consider the substitution $(12)(3456)$, or $(12)(34)(45)(56)$;

expressed to give the effect of this on $\{A\}, \{B\}, ..., \{F\}$, it has the forms

$$\binom{F}{C_0}\binom{C_0}{B}\binom{B}{E_0}\binom{E_0}{A}, \quad \binom{D}{B_0}\binom{B_0}{C}\binom{C}{F_0}\binom{F_0}{B}, \quad \binom{A}{E_0}\binom{E_0}{A}\binom{A}{D_0}\binom{D_0}{C},$$

$$\binom{B}{D_0}\binom{D_0}{F}\binom{F}{C_0}\binom{C_0}{D}, \quad \binom{C}{F_0}\binom{F_0}{D}\binom{D}{A_0}\binom{A_0}{E}, \quad \binom{E}{A_0}\binom{A_0}{E}\binom{E}{B_0}\binom{B_0}{F},$$

so that we may write

$$(12)(3456) = (BD)(AFEC);$$

this again leaves none of $\{A\}, ..., \{F\}$ unaltered; and the same is therefore also true of the ninety even substitutions of $1, 2, ..., 6$, of this form.

There remain then, altogether, $360 - 40 - 90$ or 230 even substitutions arising in the six groups $(A, A_0) ..., (F, F_0)$ which leave respectively $\{A\}, ..., \{F\}$ unaltered. This is verified by examination of these groups in detail: taking these groups in order, and omitting at any stage the substitutions which have already appeared, it is found that these groups use respectively

$$60, \quad 48, \quad 39, \quad 32, \quad 27, \quad 24$$

substitutions, whose total tale is 230.

In particular, it may be interesting to enumerate the cyclical substitutions, arising respectively in $(A, A_0), ..., (F, F_0)$; which are all different. In this enumeration, the six substitutions put down in the first row, and their first four powers, give twenty-four substitutions belonging to (A, A_0), and so on. The substitutions in any row are arranged so that they leave unaltered respectively the numbers $1, 2, ..., 6$.

$(A, A_0).(26435), \quad (15436), \quad (14265), \quad (13256), \quad (16234), \quad (14532),$

$(B, B_0).(23465), \quad (15463), \quad (14562), \quad (16253), \quad (13264), \quad (14235),$

$(C, C_0).(23546), \quad (13456), \quad (12465), \quad (15236), \quad (16432), \quad (12534),$

$(D, D_0).(26534), \quad (14536), \quad (15264), \quad (16235), \quad (13246), \quad (15432),$

$(E, E_0).(25436), \quad (16435), \quad (14256), \quad (13265), \quad (14632), \quad (15234),$

$(F, F_0).(25364), \quad (14365), \quad (15426), \quad (12356), \quad (16324), \quad (14523).$

With $\vartheta = vwu$, $\phi = vwu^2$, as before, those in the first row, here, are respectively

$$\phi^2\vartheta\phi^{-2}, \quad \phi^4\vartheta\phi^{-4}, \quad \vartheta, \quad \phi\vartheta\phi^{-1}, \quad \phi, \quad \phi^3\vartheta\phi^{-3},$$

and, in terms of $\{A\}, \ldots, \{F\}$, we find

$$\vartheta = (BCDEF), \quad \phi = (BCDFE).$$

Those in the succeeding rows are obtainable from those in the first row by the respective transpositions

$$(36), \quad (24), \quad (45), \quad (56), \quad (23),$$

with a proper reordering of those in the derived row.

We may study the group of substitutions which leave $\{A\}$ unaltered, in a different way; namely, by considering the effect of any substitution of $x_1, x_2, \ldots, x_6$ upon

$$\xi_1 = [A_0 B], \quad \xi_2 = [A_0 C], \quad \xi_3 = [A_0 D],$$
$$\xi_4 = [A_0 E], \quad \xi_5 = [A_0 F],$$

which, equated to zero, are the primes of the pentahedron $\{A\}$. In particular, the substitutions

$$u = (123)(456), \quad v = (13)(45), \quad w = (25)(36)$$

lead to the replacement of $\xi_1, \xi_2, \ldots, \xi_5$ respectively by

$$\epsilon^2\xi_1, \ \epsilon\xi_2, \ \xi_4, \ \epsilon\xi_5, \ \epsilon^2\xi_3; \quad \epsilon^2\xi_2, \ \epsilon\xi_1, \ \xi_3, \ \epsilon\xi_5, \ \epsilon^2\xi_4;$$
$$\xi_1, \ \xi_3, \ \xi_2, \ \xi_5, \ \xi_4;$$

if we neglect the powers of ϵ entering here as factors (which, we notice, in passing, are such that the product $\xi_1\xi_2\xi_3\xi_4\xi_5$ remains unaltered), these replacements are substitutions for $\xi_1, \xi_2, \ldots, \xi_5$ which we may denote respectively by (345), $(12)(45)$, $(23)(45)$; these, which are precisely of the same forms as those we have found, expressed as substitutions for B, C, D, E, F, arising as consequences of u, v, w, generate the alternate group of sixty substitutions of $\xi_1, \xi_2, \ldots, \xi_5$; as we have seen.

We have obtained six groups of even substitutions of $1, 2, \ldots, 6$, of order 60, namely $(A, A_0), \ldots, (F, F_0)$. These imply the existence of six subgroups of order 120, of the general symmetric group, of order 720, of $1, 2, ..., 6$. It is obvious that, of the group of $n!$ substitutions of n numbers, there are n subgroups, of order $(n-1)!$, each keeping one of the n numbers unaltered. The remark that, of the symmetric group of *six* numbers, besides six such subgroups, there are six subgroups of this order which are (doubly) transitive, seems to be due to Cauchy. The composition of such a subgroup is considered by J. A. Serret, *Cours d'algèbre supérieure* (4th ed., 1879) t. ii, pp. 335–40, in close connexion with his exposition of a paper by Cauchy, *Journ. de l'École Polytechnique* (x. Cahier). The possibility of such a subgroup is referred to by Burkhardt, *Math. Ann.* XXXVIII, 1890, p. 204. Cauchy's result is proved by Burnside, *Theory of Groups*, 2nd ed., 1911, p. 208, with reference to the properties of substitutions of five numbers. It is proved by Bianchi, *Lezioni sulla teoria dei gruppi di sostituzioni*, 1899, p. 68, that any simple group of order 60 is isomorphic with the alternate group of five numbers.

Serret generates the subgroup of order 120, which he considers, as the aggregate

$$(1 + T + ... + T^4)(1 + U + ... + U^3)(1 + S + ... + S^5),$$

wherein $T = (15342)$, $U = (1234)$, $S = (152346)$. It follows, interchanging the numbers 4 and 6 in his component substitutions, that the group of sixty even substitutions which we have denoted by (A, A_0), consists of the aggregate

$$\tau^p \omega^q \sigma^r \begin{pmatrix} p = 0, ..., 4; \quad q = 0, ..., 3; \quad r = 0, ..., 5 \\ q + r \text{ even} \end{pmatrix},$$

in which $\tau = (15362)$, $\omega = (1236)$, $\sigma = (152364)$.

The other five subgroups, $(B, B_0), ..., (F, F_0)$, are obtainable from this as we have explained. In regard to these subgroups reference should also be made to Todd, *Proc. Camb. Phil. Soc.* XLI, 1945, pp. 66–8 (dated 7 July 1944).

(21) **The transformation of the family {A} by means of Burkhardt's transformations.** It is interesting to compare the results we have obtained for the change of the pentahedron

$\{A\}$ into itself by means of substitutions of $x_1, \ldots, x_6$ only, with the results previously (§(18)) obtained for the change of $\{AB\}$ into itself by means of Burkhardt's transformations B, D, E, F, where $E = (DS^2)^2$, and $F = (DC^2)^2$.

We have $p(AB_0)\{AB\} = \{AB\}$, and $p(B_0C)\{AB\} = \{A\}$, as the geometrical interpretation of the projections shews. We put $\psi = p(AB_0)\,p(B_0C)$, which proves to be given by

$$x_2' = x_4 + \epsilon x_5 + \epsilon x_6, \quad -x_6' = x_1 + \epsilon^2 x_2 + \epsilon^2 x_3,$$
$$x_3' = \epsilon x_4 + x_5 + \epsilon x_6, \quad -x_4' = \epsilon^2 x_1 + x_2 + \epsilon^2 x_3,$$
$$x_1' = \epsilon x_4 + \epsilon x_5 + x_6, \quad -x_5' = \epsilon^2 x_1 + \epsilon^2 x_2 + x_3,$$

or by $(y_0', y_1', \ldots, y_4') = (y_0, \epsilon y_2, \epsilon^2 y_1, y_4, y_3),$

and is such that $\psi^2 = 1$, so that ψ is equally $p(B_0C)\,p(AB_0)$. We have used, §(3), the transformation χ, given by

$$x_1' = x_1 + \epsilon x_2 + \epsilon x_3, \quad -x_4' = x_4 + \epsilon^2 x_5 + \epsilon^2 x_6,$$
$$x_2' = \epsilon x_1 + x_2 + \epsilon x_3, \quad -x_5' = \epsilon^2 x_4 + x_5 + \epsilon^2 x_6,$$
$$x_3' = \epsilon x_1 + \epsilon x_2 + x_3, \quad -x_6' = \epsilon^2 x_4 + \epsilon^2 x_5 + x_6,$$

or by $(y_0', y_1', \ldots, y_4') = (y_0, \epsilon^2 y_1, \epsilon y_2, \epsilon y_3, \epsilon^2 y_4),$

and we find that

$$\psi = \chi.p(16)\,p(24)\,p(35) = \chi.p(AC_0)\,p(A_0C).$$

The transformation ψ is such that $\psi\{AB\} = \{A\}$. Hence, the transformations leaving $\{A\}$ unaltered which correspond to Burkhardt's transformations B, D, E, F for $\{AB\}$, are

$$B_1 = \psi B\psi, \quad D_1 = \psi D\psi, \quad E_1 = \psi E\psi, \quad F_1 = \psi F\psi.$$

It can be found by computation that

$$B_1 = p(AD_0)\,p(AF_0)\,p(25)\,p(36),$$

and is given by

$$x_1' = x_3 + \epsilon^2 x_2 + \epsilon^2 x_6, \quad -x_6' = x_1 + \epsilon x_4 + \epsilon x_5,$$
$$x_2' = \epsilon^2 x_3 + x_2 + \epsilon^2 x_6, \quad -x_3' = \epsilon x_1 + x_4 + \epsilon x_5,$$
$$x_4' = \epsilon^2 x_3 + \epsilon^2 x_2 + x_6, \quad -x_5' = \epsilon x_1 + \epsilon x_4 + x_5,$$

which lead to

$$x_4' - \epsilon^2 x_6' = (\epsilon^2 - 1)(x_1 + x_2 + x_3), \quad x_1' + x_2' + x_3' = (\epsilon - 1)(x_4 - \epsilon^2 x_6).$$

Further it can be shewn that

$$D_1 = p(AC_0), \quad E_1 = (x_1 x_2 x_3)(x_4 x_5 x_6), \quad F_1 = (x_1 x_3)(x_4 x_5),$$

so that E_1 and F_1 are precisely those denoted by u and v in the preceding section, §(20), p. 87.

Putting then

$$\xi_1 = [14.25.36], \quad \xi_2 = [16.35.24], \quad \xi_3 = [13.26..45],$$

$$\xi_4 = [12.34.56], \quad \xi_5 = [15.46.23],$$

the transformed values of these, under the transformations, are found to be given by

	B_1	E_1	F_1	D_1
ξ_1'	ξ_1	$\epsilon^2 \xi_1$	$\epsilon^2 \xi_2$	ξ_1
ξ_2'	ξ_3	$\epsilon \xi_2$	$\epsilon \xi_1$	$-\xi_2$
ξ_3'	$-\xi_2$	ξ_4	ξ_3	ξ_3
ξ_4'	ξ_5	$\epsilon \xi_5$	$\epsilon \xi_5$	ξ_4
ξ_5'	$-\xi_4$	$\epsilon^2 \xi_3$	$\epsilon^2 \xi_4$	ξ_5

The projections $p(AD_0)$, $p(AF_0)$, which are commutable, have the respective effects upon $\xi_1, ..., \xi_5$ of merely changing the signs of ξ_3 and ξ_5 (§(7)); the transformation $p(AF_0) p(AD_0) B_1$, or $p(25) p(36)$, or $(25)(36)$, is that denoted by w in the preceding article, §(20), and has the effect there noted. We may notice also that the product wvu, or $p(AD_0) p(AF_0) B_1 F_1 E_1$, produces in $\xi_1, ..., \xi_5$ the cyclical change $(\xi_1 \xi_2 \xi_4 \xi_5 \xi_3)$, which clearly leaves unaltered the equation of the Burkhardt primal in $\xi_1, \xi_2, ..., \xi_5$ found in §(15). We note, too, besides $\psi\{AB\} = \{A\}$, $\psi\{A\} = \{AB\}$, that $\psi\{B_0\} = \{B_0\}$; and that every one of the families $\{A\}$, $\{AB\}$, $\{B_0\}$ is changed into itself by both the transformations B and B_1.

(22) **Derivation of the Burkhardt primal from a quadric.** Consider a quadric, in space of four dimensions, passing through the angular points of the simplex of reference, so that its equation does not contain the squares of the coordinates

$X_1, X_2, ..., X_5$, but contains ten terms. Suppose that the equation of this quadric is unaltered by the transformations B'_1, F'_1, E'_1, given by

	B'_1	F'_1	E'_1
X'_1	X_1	ϵX_2	ϵX_1
X'_2	X_3	$\epsilon^2 X_1$	$\epsilon^2 X_2$
X'_3	X_2	X_3	X_4
X'_4	X_5	$\epsilon^2 X_5$	$\epsilon^2 X_5$
X'_5	X_4	ϵX_4	ϵX_3

These are such that $B'_1 F'_1 E'_1$ leads to $(X'_1, ..., X'_5) = (X_2 X_4 X_1 X_5 X_3)$, that is, to the cyclical transformation $(X_1 X_2 X_4 X_5 X_3)$. It is found on trial that this condition determines the quadric, and that its equation has the form

$$X_1 X_2 + X_2 X_4 + X_4 X_5 + X_5 X_3 + X_3 X_1$$
$$+ \epsilon(X_1 X_4 + X_2 X_5 + X_4 X_3 + X_5 X_1 + X_3 X_2) = 0.$$

The tangent primes of the quadric at the angular points of the simplex of reference are $\overline{X}_1 = 0, ..., \overline{X}_5 = 0$, where

$$\tfrac{1}{2}\epsilon^2 (\overline{X}_1, \overline{X}_2, ..., \overline{X}_5) \begin{pmatrix} 0, & \epsilon^2, & \epsilon^2, & 1, & 1 \\ \epsilon^2, & 0, & 1, & \epsilon^2, & 1 \\ \epsilon^2, & 1, & 0, & 1, & \epsilon^2 \\ 1, & \epsilon^2, & 1, & 0, & \epsilon^2 \\ 1, & 1, & \epsilon^2, & \epsilon^2, & 0 \end{pmatrix} \begin{aligned} & (X_1, X_2, ..., X_5), \\ & = \Omega_0(X_1, X_2, ..., X_5), \\ & \qquad\qquad \text{say.} \end{aligned}$$

These tangent primes form a simplex whose angular points are respectively given by the rows of

$$\tfrac{1}{2} \begin{pmatrix} 0, & \epsilon, & \epsilon, & 1, & 1 \\ \epsilon, & 0, & 1, & \epsilon, & 1 \\ \epsilon, & 1, & 0, & 1, & \epsilon \\ 1, & \epsilon, & 1, & 0, & \epsilon \\ 1, & 1, & \epsilon, & \epsilon, & 0 \end{pmatrix}, = \Omega, \text{ say,}$$

and the matrices are such that $\Omega \Omega_0 = 1$. Thus the angular points of the simplex of reference, and of the simplex formed by the tangent primes, respectively lie on the prime faces each of the other simplex; and the quadric touches any prime face of either simplex at an angular point of the other simplex.

Save for the multiplying powers of ϵ, the substitutions of $X_1, \ldots, X_5$ effected by B_1', F_1' and E_1', are respectively $(X_2 X_3)(X_4 X_5)$, $(X_1 X_2)(X_4 X_5)$ and $(X_3 X_4 X_5)$, or say $(23)(45)$, $(12)(45)$ and (345), which we have called in §(20), p. 83, respectively w, v, u. It follows then, by what we have said, that the quadric is subject to, and defined by, a group of sixty even linear substitutions, isomorphic with the alternate group of five numbers.

Referring now to the equation of the Burkhardt primal given above in §(15), we see that the primal is obtainable from the quadric by the substitutions

$$X_1 = \xi_1^2, \quad X_2 = \xi_2^2, \quad \ldots, \quad X_5 = \xi_5^2,$$

or, in a usual phraseology, the quadric *represents* an involution of sets of sixteen points lying on the primal, of which any set is given by $(\pm \xi_1, \pm \xi_2, \ldots, \pm \xi_5)$. The transformations B_1', F_1', E_1' are those thence arising from the B_1, F_1, E_1 employed in §(21). We know that any general quadric in fourfold space can be referred to a simplex to which it is both inscribed and circumscribed, as here (see for instance, Baker, *Annali di Mat.*, XVI (1937), and the interesting paper by B. Segre there referred to). The Burkhardt primal thus arises in a simple way from any general quadric, referred to such a simplex. The forty-five nodes of the primal arise, in an easily recognized exceptional way, from the angular points of the two simplexes in the space $(X_1, \ldots, X_5)$.

Remark I. The equations which connect $x_1, \ldots, x_6$ with $\xi_1, \ldots, \xi_5$ have been put down in detail in §(15). Let $\bar{\xi}_1$ denote the conjugate imaginary of ξ_1; or, ξ_1 being $[14.25.36]$, let

$$\bar{\xi}_1 = \lfloor 14.36.25 \rfloor;$$

and similarly for $\bar{\xi}_2, \ldots, \bar{\xi}_5$. Thus $\xi_1 = 0$ is the prime face of the pentahedron $\{A\}$ which is opposite to the angular point $(A B_0)$, and $\bar{\xi}_1 = 0$ is the tangent prime of the quadric $x_1^2 + \ldots + x_6^2 = 0$ at this point. We find then that

$$-\epsilon^2(\bar{\xi}_1, \bar{\xi}_2, ..., \bar{\xi}_5) = \Omega(\xi_1, \xi_2, ..., \xi_5).$$

If we define Q by $\qquad Q = \xi_1\bar{\xi}_1 + ... + \xi_5\bar{\xi}_5,$

this leads to

$$Q = 6(x_1^2 + ... + x_6^2) - (x_1 + ... + x_6)^2 = 6(x_1^2 + ... + x_6^2),$$

and to

$$-\epsilon Q = \xi_1\xi_2 + \xi_2\xi_4 + \xi_4\xi_5 + \xi_5\xi_3 + \xi_3\xi_1$$
$$+ \epsilon^2(\xi_1\xi_4 + \xi_2\xi_5 + \xi_4\xi_3 + \xi_5\xi_1 + \xi_3\xi_2);$$

thus the quadric in $X_1, ..., X_5$, considered above, differs in form from the expression of $x_1^2 + ... + x_6^2$, in terms of $\xi_1, \xi_2, ..., \xi_5$, only by the change of ϵ into ϵ^2. Thus the quadric $x_1^2 + ... + x_6^2 = 0$, which arose in connexion with the double six of pentahedra, in § (16), when expressed by $\xi_1, ..., \xi_5$, may be defined as allowing linear substitutions in these coordinates, which differ from B_1', F_1', E_1', above, only by change of ϵ into ϵ^2.

Further, putting $\xi_1, \xi_2, \xi_3, \xi_4, \xi_5$ respectively equal to $\epsilon x, \epsilon y, z, t, u$, as in § (15), the quadric $x_1^2 + ... + x_6^2 = 0$, in these coordinates, becomes

$$yz + xt + \epsilon(zx + yt) + \epsilon^2(xy + zt) + u(x + y + z + t) = 0;$$

and, we find for the quadric $x_1 x_4 + x_2 x_5 + x_3 x_6 = 0$, of § (16), the equation in these coordinates,

$$yz - xt + \epsilon(-zx + yt) + \epsilon^2(-xy + zt) + u(-x + y + z + t) = 0,$$

differing from the former only by the sign of x (cf. § (16)); and, for the quadric

$$x_2 x_3 + x_3 x_1 + x_1 x_2 + \epsilon(x_5 x_6 + x_6 x_4 + x_4 x_5) = 0,$$

the equation in these coordinates

$$-(yz + xt) - \epsilon(zx + yt) + \epsilon^2(xy + zt) + u(-x - y + z + t) = 0,$$

differing from Q only by the change of sign of x and y.

Remark II. The quadric surface

$$a(yz + xt) + b(zx + yt) + c(xy + zt) = 0,$$

where $a+b+c = 0$, is a cone with vertex at $(1,1,1,1)$, the four generators from this to the angular points of the tetrahedron (x, y, z, t) having a cross-ratio, on the cone, equal to the negative of one of the six ratios of two of a, b, c. These generators are then 'equi-anharmonic' when $a, b, c = 1, \epsilon, \epsilon^2$, in any order. As all the pentahedra of the Burkhardt primal are similar, it follows that the equation of the primal, for any pentahedron, agrees with that marked (II) in § (15), save for unessential multipliers of the prime faces. In particular, for the pentahedron $\{B_0\}$, it will be found that this is so for

$$x = \epsilon[14.25.36], \quad y = [13.56.24], \quad z = [12.46.35],$$
$$t = \epsilon[15.34.26], \quad u = \epsilon^2[16.23.45].$$

From each of the pentahedra, there is an involution of sixteen points on the primal; and these are thus all representable by the same quadric; which itself allows sixty self-transformations (and we notice that $27.16.60 = 2^3.3^4.40$).

Postscript to p. 79. In a way used by Dr Todd the whole group can be formed from the A of p. 78 and the ϑ of p. 87, which is $p(14)\,p(42)\,p(26)\,p(65)$. For these give $A^3 = p(14)$ and

$$\vartheta^r p(14)\vartheta^{-r} = p(24),\ p(26),\ p(56),\ p(15);\ (r = 1, 2, 3, 4),$$

while $\qquad\qquad A^{-1}p(56)A = p(A_0B).$

Whence, with A and ϑ we can form the projections from the five nodes (14), (24), (26), (15), (A_0B). From the first four of these, by κ-lines, we can obtain the other six of the ten nodes (ij), with $i, j \neq 3$. Also the node (A_0B) can be joined to any other of the 30 nodes (PQ_0) by an open polygon of κ-lines, each containing one of the ten nodes (ij). Each of the other five nodes is an angular point of a Jordan polyhedron of which four angular points are constructed.

APPENDIX

Note 1. *The generation of desmic systems of tetrahedra in ordinary space*

The notion of three tetrahedra, in space of three dimensions, of which any two are in perspective with one another from every angular point of the third, has occurred frequently in what precedes, especially in the theory of κ-lines. It is familiar that, with any three such tetrahedra, are associated three other tetrahedra, also forming a desmic system; the edges of the three latter tetrahedra are the same lines as the edges of the former. We note here a method in which the six tetrahedra arise together; it appears that the six tetrahedra arise from a group of six linear transformations, which is isomorphic with the symmetric group of substitutions of three objects.

It is well known that an orthogonal matrix, of determinant unity, can be expressed by the ratios of four independent variables ξ, η, ζ, τ, in the form

$$\begin{pmatrix} a_1, & b_1, & c_1 \\ a_2, & b_2, & c_2 \\ a_3, & b_3, & c_3 \end{pmatrix}, \text{ save for the factor } (\tau^2 + \xi^2 + \eta^2 + \zeta^2)^{-1},$$

$$= \begin{pmatrix} \tau^2 + \xi^2 - \eta^2 - \zeta^2, & 2(\xi\eta - \zeta\tau), & 2(\xi\zeta + \eta\tau) \\ 2(\eta\xi + \zeta\tau), & \tau^2 + \eta^2 - \zeta^2 - \xi^2, & 2(\eta\zeta - \xi\tau) \\ 2(\zeta\xi - \eta\tau), & 2(\zeta\eta + \xi\tau), & \tau^2 + \zeta^2 - \xi^2 - \eta^2 \end{pmatrix},$$

the ratios of τ, ξ, η, ζ being conversely determinable in terms of the elements of the orthogonal matrix by the equations

$$\tau^{-1}\xi = q/r' = r/q', \quad \tau^{-1}\eta = r/p' = p/r', \quad \tau^{-1}\zeta = p/q' = q/p',$$

where p, q, r, p', q', r' are given by

$$p = b_3 + c_2, \quad q = c_1 + a_3, \quad r = a_2 + b_1,$$
$$p' = b_3 - c_2, \quad q' = c_1 - a_3, \quad r' = a_2 - b_1.$$

We may investigate then what are the transformations of ξ, η, ζ, τ which result from the interchanges of the rows of the orthogonal matrix. It is sufficient for this purpose to find the changes in ξ, η, ζ, τ due to the three operations: (1) interchange of the second and third rows of the orthogonal matrix, with a subsequent change of sign of all rows; (2) interchange of the third and first rows of this matrix; (3) interchange of the first and second rows, with similar changes of sign throughout.

It is found that these interchanges are effected by three linear transformations of ξ, η, ζ, τ respectively given by

$$(1) \quad (\xi', \eta', \zeta', \tau') = (\eta + \zeta, \qquad \tau - \xi, -\tau - \xi, \zeta - \eta),$$
$$(2) \quad (\xi', \eta', \zeta', \tau') = (-\tau - \eta, \quad \zeta + \xi, \quad \tau - \eta, \xi - \zeta),$$
$$(3) \quad (\xi', \eta', \zeta', \tau') = (\tau - \zeta, \quad -\tau - \zeta, \quad \xi + \eta, \eta - \xi).$$

These linear transformations we denote by

$$(\xi', \eta', \zeta', \tau') = \phi_1(\xi, \eta, \zeta, \tau),$$
$$(\xi', \eta', \zeta', \tau') = \phi_2(\xi, \eta, \zeta, \tau), (\xi', \eta', \zeta', \tau') = \phi_3(\xi, \eta, \zeta, \tau),$$

so that

$$\phi_1 = \begin{pmatrix} 0, & 1, & 1, & 0 \\ -1, & 0, & 0, & 1 \\ -1, & 0, & 0, & -1 \\ 0, & -1, & 1, & 0 \end{pmatrix}, \quad \phi_2 = \begin{pmatrix} 0, & -1, & 0, & -1 \\ 1, & 0, & 1, & 0 \\ 0, & -1, & 0, & 1 \\ 1, & 0, & -1, & 0 \end{pmatrix},$$

$$\phi_3 = \begin{pmatrix} 0, & 0, & -1, & 1 \\ 0, & 0, & -1, & -1 \\ 1, & 1, & 0, & 0 \\ -1, & 1, & 0, & 0 \end{pmatrix}.$$

These are such that the six operations 1, $(\phi_2 \phi_3)$, $(\phi_2 \phi_3)^{-1}$, ϕ_1, ϕ_2, ϕ_3 form a group, which is isomorphic with the group of substitutions of three objects. For we find

$$\phi_2 \phi_3 = \phi_3 \phi_1 = \phi_1 \phi_2 = \begin{pmatrix} 1, & -1, & 1, & 1 \\ 1, & 1, & -1, & 1 \\ -1, & 1, & 1, & 1 \\ -1, & -1, & -1, & 1 \end{pmatrix},$$

$$\phi_3\phi_2 = \phi_1\phi_3 = \phi_2\phi_1 = \begin{pmatrix} 1, & 1, & -1, & -1 \\ -1, & 1, & 1, & -1 \\ 1, & -1, & 1, & -1 \\ 1, & 1, & 1, & 1 \end{pmatrix},$$

$$\phi_3\phi_2 = 4(\phi_2\phi_3)^{-1}, \quad (\phi_2\phi_3)^2 = -2(\phi_3\phi_2),$$
$$(\phi_2\phi_3)^3 = -8, \quad \phi_1\phi_2\phi_3 = -2\phi_2, \quad \phi_1\phi_3\phi_2 = -2\phi_3.$$

If now we take any four points A, B, C, D, which are the angular points of a tetrahedron, and put

$$(X_1, Y_1, Z_1, T_1) = \phi_1(A, B, C, D)$$
$$= (B+C, -A+D, -A-D, -B+C);$$
$$(X_2, Y_2, Z_2, T_2) = \phi_2(A, B, C, D)$$
$$= (-B-D, C+A, -B+D, -C+A);$$
$$(X_3, Y_3, Z_3, T_3) = \phi_3(A, B, C, D)$$
$$= (-C+D, -C-D, A+B, -A+B),$$

then (X_1, Y_1, Z_1, T_1), (X_2, Y_2, Z_2, T_2), (X_3, Y_3, Z_3, T_3) are the angular points of three desmic tetrahedra, lying in threes on sixteen lines; we have for instance

$$X_1 + X_2 + X_3 = 0, \quad X_1 - Y_2 - T_3 = 0,$$
$$X_1 + Z_2 + Y_3 = 0, \quad X_1 + T_2 - Z_3 = 0.$$

Also, if we put

$$(A', B', C', D') = \phi_2\phi_3(A, B, C, D) = (A-B+C+D,$$
$$A+B-C+D, -A+B+C+D, -A-B-C+D);$$
$$(A'', B'', C'', D'') = \phi_3\phi_2(A, B, C, D) = (A+B-C-D,$$
$$-A+B+C-D, A-B+C-D, A+B+C+D),$$

then (A, B, C, D), (A', B', C', D'), (A'', B'', C'', D''), or, essentially, $A, B, C, D)$, $\phi_2\phi_3(A, B, C, D)$, $(\phi_2\phi_3)^2(A, B, C, D)$ are the angular points of three desmic tetrahedra, lying in threes on sixteen lines, of which, for instance, four lines are

$$AA'A'', \quad AB'C'', \quad AC'D'', \quad AD'B''.$$

Any one of the six tetrahedra may be regarded as fundamental. For instance

$$(A, B, C, D) = -\tfrac{1}{2}\phi_1(X_1, Y_1, Z_1, T_1),$$
$$(A', B', C', D') = \phi_3(X_1, Y_1, Z_1, T_1),$$
$$(A'', B'', C'', D'') = \phi_2(X_1, Y_1, Z_1, T_1),$$

while, also,

$$(X_2, Y_2, Z_2, T_2) = -\tfrac{1}{2}\phi_3\,\phi_2(X_1, Y_1, Z_1, T_1),$$
$$(X_3, Y_3, Z_3, T_3) = -\tfrac{1}{2}\phi_2\,\phi_3(X_1, Y_1, Z_1, T_1).$$

Again, we have $Y_1 + Z_1 = -2A$, $Y_1 - Z_1 = 2D$, so that the points Y_1, Z_1 are on the edge AD, and are harmonic conjugates in regard to A and D, etc.; and the eighteen edges of the tetrahedra (X_1, Y_1, Z_1, T_1), (X_2, Y_2, Z_2, T_2), (X_3, Y_3, Z_3, T_3), properly taken, are the edges of the tetrahedra

$$(A, B, C, D), \quad (A', B', C', D'), \quad (A'', B'', C'', D'').$$

Also we have

$$A' + B' = -B'' + D'', \quad -A' + B' = 2(B - C), \quad B'' + D'' = 2(B + C),$$

so that the edges BC, $A'B'$, $B''D''$ lie in a plane, these being the diagonals of the quadrilateral formed by the lines

$$A'D''B, \quad A'CB'', \quad B'D''C, \quad B'BB'', \text{ etc.}$$

Note 2.   *On the group of substitutions of the lines of a cubic surface in ordinary space*

The theory given above in this volume deals essentially with the group of the substitutions of the tritangent planes of a cubic surface, which may be said to have been the group studied by Jordan. But the lines are curves of the cubic surface; and may therefore, from the point of view of coresiduation on the surface, be regarded as linearly expressible in terms of seven such curves. These we may take to be six mutually skew lines $a_1, a_2, \ldots, a_6$, together with a cubic curve u, not meeting any of these lines

(*Proc. Lond. Math. Soc.* XI (1912), p. 298). The group is then representable by linear transformations for these seven, regarded as algebraic variables; the other lines, in Schläfli's notation, are then represented by $b_i = 2u - s + a_i$, $c_{ij} = u - a_i - a_j$, where s denotes $a_1 + a_2 + \ldots + a_6$, and i, j have the values $1, 2, \ldots, 6$. It appears then that the group is generated by the symmetric group of substitutions of $a_1, a_2, \ldots, a_6$, combined with such transformations as interchange among themselves the seventy-two sets of six mutually skew lines, each with its associated cubic curve. (Burnside, *Proc. Lond. Math. Soc.* X (1911), pp. 299–301). These seventy-two sets are typified by

$$\alpha = (a_1, a_2, \ldots, a_6, u),$$

$$\beta = (2u - s + a_1, \ldots, 2u - s + a_6, 5u - 2s),$$

$$\nu_{12} = (a_1, 2u - s + a_1, u - a_2 - a_3, u - a_2 - a_4, \ldots,$$
$$u - a_2 - a_6, 3u - s + a_1 - a_2),$$

$$\lambda_{123} = (a_1, a_2, a_3, u - a_5 - a_6, u - a_6 - a_4, u - a_4 - a_5,$$
$$2u - a_4 - a_5 - a_6),$$

$$\mu_{123} = (u - a_2 - a_3, u - a_3 - a_1, u - a_1 - a_2, 2u - s + a_4,$$
$$2u - s + a_5, 2u - s + a_6, 4u - s - a_1 - a_2 - a_3),$$

where the associated cubic curve in any set may be immediately written down by the fact that $3u - s$ remains unaltered by any transformation of the group (as does $u^2 - a_1^2 - a_2^2 - \ldots - a_6^2$). The interchanges of these seventy-two sets are all obtainable by combining permutations of $a_1, a_2, \ldots, a_6$ among themselves with a single transformation. This single transformation may, for instance, be taken to be L, given by

$$a_1' = a_1, \qquad a_2' = 2u - s + a_1, \quad a_3' = u - a_2 - a_3,$$
$$a_4' = u - a_2 - a_4, \ldots, \quad a_6' = u - a_2 - a_6, \quad u' = 3u - s + a_1 - a_2,$$

and this transformation, as changing the set α into the set ν_{12}, may be denoted by $\begin{pmatrix} \nu_{12} \\ \alpha \end{pmatrix}$; it is such that $L^2 = 1$. That it is effective, when combined with the permutations, for the interchange of the seventy-two sets, follows from the equations

$$L\alpha = \nu_{12}, \text{ or } L\nu_{12} = \alpha, \quad L\nu_{21} = \beta, \quad L\nu_{13} = \lambda_{123}, \quad L\nu_{31} = \mu_{123},$$

which can be immediately verified.

The group so obtained contains $72(6!)$, or $2^4 . 3^4 . 40$ transformations (but it contains a subgroup of order $2^3 . 3^4 . 40$, which effects only even substitutions of the tritangent planes).

The group is sufficiently represented by supposing $u = \frac{1}{3}s$, regarding then the group as a group of transformations among $a_1, a_2, ..., a_6$ only. In general the condition for three lines to lie in a tritangent plane is that the symbols of these lines should have $3u - s$ for sum; in the curtailed group for $a_1, ..., a_6$ only, the condition will then be that the sum of these three symbols should vanish.

It is thus suggested that we should interpret the symbols of the lines as being the symbols of points in space of five dimensions, every such point being dependent upon six points; the points so arising from the lines of a tritangent plane of the cubic surface will then be upon a line. We are thus led to a configuration of twenty-seven points, in this space, lying in threes upon forty-five lines, there being five of these lines passing through each of these twenty-seven points. The group under consideration is that of the linear self-transformations of this configuration. There are forty ways in which the twenty-seven points may be divided into three batches of each nine points, those of a batch being the intersections of three skew lines with three other skew lines (so that the points of a batch lie in a threefold space); there remain then $45 - 3.6$, or twenty-seven other lines of the configuration; each of these contains one point of each of the three batches spoken of, so that three of these twenty-seven transversal lines pass through every one of the nine points of any one of the three batches. Further, among the twenty-seven points of the configuration there are seventy-two sets of six points, each set forming a simplex; none of the fifteen joining lines of two angular points of one of these simplexes is among the forty-five original lines (does not contain a third point of the original twenty-seven points). We may, for instance, borrowing Schläfli's notation, take the three batches to consist of the nine points, respectively,

$$c_{23}\,b_2\,a_3 \qquad c_{56}\,a_6\,b_5 \qquad c_{14}\,c_{36}\,c_{25}$$
$$a_2\,c_{12}\,b_1 \qquad b_6\,c_{64}\,a_4 \qquad c_{26}\,c_{15}\,c_{34}$$
$$b_3\,a_1\,c_{13} \qquad a_5\,b_4\,c_{45} \qquad c_{35}\,c_{24}\,c_{16}$$

the points b_i, c_{ij} being, in terms of $a_1, ..., a_6$, given by

$$b_i = a_i - \tfrac{1}{3}s, \qquad c_{ij} = \tfrac{1}{3}s - a_i - a_j,$$

in which $s = a_1 + a_2 + ... + a_6$. In each batch, the three points in every row of the scheme, and those in every column, are in a line; and through any point of one batch pass three lines, each containing a point from both the other batches; for instance c_{23}, c_{56}, c_{14}; c_{23}, c_{64}, c_{15}; c_{23}, c_{45}, c_{16} are three lines through c_{23}.

Moreover, we easily see that the seventy-two simplexes consist of thirty-six pairs, in which those of a pair are in perspective with one another from a point.

At such a configuration also Burkhardt arrives (*Math. Ann.* XLI (1892), p. 326) starting from transformations arising by consideration of double theta functions (but different functions from those leading to the Burkhardt primal); his acknowledgements are to Klein (*Liouville's J.*, IV, 1888; *Ges. Abh.* II, p. 473), Witting (*Math. Ann.* XXIX (1887), and Maschke (*Math. Ann.* XXXIII (1889)). But his twenty-seven points have definite numerical coordinates, those of the first batch of nine points being $(0, a, b, 0, a_0, b_0)$, where $a^3 = 1$, $b^3 = 1$ and a_0, b_0 are the conjugate imaginaries respectively of a, b; those of the second batch being $(a, 0, b, a_0, 0, b_0)$; and those of the third batch being $(a, -b, 0, a_0, -b_0, 0)$; moreover, instead of twenty-seven points, he has twenty-seven linear complexes of lines in space of three dimensions, and instead of forty-five lines he has forty-five linear congruences of lines. His result was the first exhibition of the group of the lines of a cubic surface as a linear group in six coordinates. Burnside (*Proc. Lond. Math. Soc.* X (1911), pp. 299–

301) gave an independent proof that Burkhardt's equations generate a group, and shewed that, in addition to the symmetrical group of substitutions of $a_1, a_2, ..., a_6$, only one (not two) transformation is necessary to generate the group. The group has also been considered in relation to a configuration founded on twenty-seven points in hyperspace by Coxeter (*Trans. Royal Soc.* vol. 229 (1930), p. 418; and subsequent papers). Cf. also Coble (*loc. cit.*, Introduction).

Professor B. Segre, in his recent monograph, *The Non-singular Cubic Surfaces* (Oxford, 1942), has generated the group from the thirty-six transpositions of the lines in the two rows of the double sixes of lines, without the expression of the lines in terms of seven elements. From the point of view here taken, it is sufficient, for the generation of the group, to combine five such transpositions, namely $(\nu_{61}, \nu_{16}), (\nu_{62}, \nu_{26}), ..., (\nu_{65}, \nu_{56})$, which are equivalent to the transpositions $(a_6, a_1), (a_6, a_2), ..., (a_6, a_5)$, with the single transposition $(\lambda_{456}, \mu_{456})$, which is equivalent to the transformation of the set α into the set λ_{123}, thus using six transpositions in all.

Reference should also be made to the investigation of the group of the lines of a cubic surface given in Dickson, *Linear Groups* (1901), Chap. XIV, whose notation for the tritangent planes enables him to put the group in relation with the algebraic theory of the so-called Abelian group, originally derived from the theory of the linear transformation of the periods of theta functions. If we make a linear transformation of the four variables x_1, x_2, y_1, y_2, and the same transformation of the four variables u_1, u_2, v_1, v_2, such that the function

$$u_1 y_1 + u_2 y_2 - v_1 x_1 - v_2 x_2$$

is unaltered thereby (cf. the author's *Abel's Theorem* (1897), p. 538), the sixteen coefficients in the transformation are subject only to six conditions. But, if these coefficients be integral numbers, and be reduced, modulus 3, each to 0, 1, or 2, the number of such transformations is finite, being $(3^4 - 1) . 3^3 . (3^2 - 1) . 3$, or $2^4 . 3^4 . 40$.

INDEX OF NOTATIONS

(The references are to the sections)

Printed in the United States
By Bookmasters